NEW HEALTH CONSIDERATIONS IN WATER TREATMENT

NEW HEALTH CONSIDERATIONS IN WATER TREATMENT

Edited by

ROGER HOLDSWORTH, LRSC, MIWEM (Dip), DMS
Water Training *international*

Based on papers presented at a symposium held on 21 November 1990
organised by the Tyne and Humber Branch of
The Institution of Water and Environmental Management
and sponsored by Water Training *international*

Published in Association with
The Institution of Water and Environmental Management

Avebury Technical

Aldershot · Brookfield USA · Hong Kong · Singapore · Sydney

© Roger Holdsworth 1991

All rights reserved. No part of this publication may be reproduced, stored in a retrieval system, or transmitted in any form or by any means, electronic, mechanical, photocopying, recording or otherwise without the prior permission of the publisher.

Published by
Avebury Technical
Academic Publishing Group
Gower House
Croft Road,
Aldershot
Hants
GU11 3HR
England

Gower Publishing Company
Old Post Road
Brookfield
Vermont 05036
USA

A CIP catalogue record for this book is available from the British Library and the US Library of Congress.

ISBN 1 85628 803 1

Printed and bound in Great Britain by
Billing and Sons Ltd, Worcester

Contents

Directory of contributors   viii

Editor's preface   ix

1. **Water Quality - Does 'Don't Know' mean 'Safe'?**   1
    J.W. Bridges, D. Wheeler and O. Bridges
    - Historic aspects   1
    - Chemicals of particular concern   4
    - Identification of particularly at risk groups of the population   7
    - Exposure to mixtures of chemicals   7
    - Future prospects   11

2. **Developments in Health Related Quality Standards for Chemicals in Drinking Water**   13
    J.K. Fawell
    - Introduction   13
    - Approaches to setting standards   14
    - The scientific basis of standards   14
    - Additional assumptions in setting standards   16
    - Some examples of uncertainties in setting standards   18
    - Discussion   20

3. **Recently Recognised Concerns in Drinking Water Microbiology**   23
    J. Watkins and S. Cameron
    - Introduction   23
    - Waterborne disease   23
    - The faecal indicator   25
    - Microbial pathogens   26
    - Surface water quality   32
    - Discussion   32

4. **Advanced Treatments for Better Drinking Water Quality**   **39**
   J. Mallevialle and *J.P. Duguet*
   - Introduction   39
   - Microbiological parameters   39
   - Nitrates   41
   - Pesticides   41
   - Tastes and odours   42
   - Economic aspects   43
   - Conclusion   45

5. **Lessons Learnt from Waterborne Outbreaks of Cryptosporidiosis**   **47**
   C. Benton
   - Introduction   47
   - Evidence for waterborne transmission   48
   - Documented waterborne outbreaks   49
   - Lessons learnt   54
   - Conclusions   57

6. **Cyanobacterial Toxins and their Significance in United Kingdom and European Waters**   **61**
   L.A. Lawton and *G.A. Codd*
   - Abstract   61
   - Introduction   61
   - Properties of cyanobacterial toxins   62
   - Occurrence of toxic blooms and cyanobacterial toxins in European waters   64
   - Events in 1989   66

7. **New Treatments for Pesticides and Chlorinated Organics Control**   **73**
D.M. Foster, A.J. Rachwal and S.L. White
    Abstract   73
    Introduction   73
    The pesticide 'problem'   75
    Sources and standards for other chlorinated organic
        compounds   77
    Pesticide and solvent levels in source waters   77
    Effects of conventional treatment processes   82
    Trihalomethane formation in conventional treatment   84
    Options for advanced water treatment   86
    Ozonation processes   87
    Activated carbon adsorption   89
    Ozone with GAC: biological activated carbon   91
    Removal of chlorinated solvents with GAC   93
    Air stripping   93
    Advanced oxidation processes   94
    Membrane processes   94
    Microbiological processes   95
    Novel absorbents   95
    Conclusions   95

Directory of contributors

DR. C. BENTON, Principal Microbiologist, Water Department, Strathclyde Regional Council, Glasgow, Scotland.
PROF. J.W. BRIDGES, Director and Professor of Toxicology, Robens Institute of Health and Safety, University of Surrey, Guildford, England.
DR. O. BRIDGES, Fellow in European Studies, University of Surrey, Guildford, England.
S. CAMERON, Virologist, Yorkshire Water Enterprises Ltd., York, England.
PROF. G.A. CODD, Department of Biological Sciences, The University of Dundee, Scotland.
DR. ING. J.P. DUGUET, Head of Drinking Water Research, Lyonnaise des Eaux, Le Pecq, France.
J.K. FAWELL, Principal Toxicologist, WRc, Medmenham, England.
D.M. FOSTER, Principal Project Engineer, Group Research and Development, Thames Water plc, Reading, England.
L.A. LAWTON, Research Assistant, Department of Biological Sciences, The University of Dundee, Dundee, Scotland.
DR. ING. J. MALLEVIALLE, Deputy Director of Research, Lyonnaise des Eaux, Le Pecq, France.
A.J. RACHWAL, Water and Treatment Distribution, Project Manager, Group Research and Development, Thames Water plc, Reading, England.
DR. J. WATKINS, Senior Microbiologist, Yorkshire Water Enterprise Ltd., York, England.
DR. D. WHEELER, Manager, Environmental Services, Robens Institute of Health and Safety, University of Surrey, Guildford, England.
S.L. WHITE, Principal Scientist, Drinking Water, Drinking Water Policy Group, Water and Environmental Science, Thames Water plc, Reading, England.

Editor's preface

The germ of the idea which culminated in the production of this book goes back to a meeting of the IWEM Tyne and Humber Branch programme sub-committee some time in the autumn of 1989. Our theme for the year's programme being planned was water quality. Within moments we realised that there were numerous health related water quality issues of serious concern to those whose work relates to the production of drinking water, which were also the subject of widespread comment and conjecture among the public at large. And so the idea was born, to present a symposium entitled 'New Health Considerations in Water Treatment'. Originally intended as a half day meeting in Spring 1990, the idea grew and the actual event became a whole day symposium held on November 21st 1990 under the chairmanship of the President of the Institution, Mr. Brian Rofe.

The symposium was planned to commence with a challenging view of water quality issues of the day by a speaker of international repute who would speak as a professional observer and focus on the real uncertainties in our knowledge of water quality. This role was ably filled by our keynote speaker, Professor Jim Bridges, Director and Professor of Toxicology at the Robens Institute of Health and Safety, University of Surrey. Co-authors of his paper are Dr. David Wheeler and Dr. O. Bridges, also of the Robens Institute.

The following two papers concentrated on chemical and microbiological concerns and were given by John Fawell who, as Principal Toxicologist at WRc, is working in the international team concerned with revision of the WHO guidelines for drinking water quality and John Watkins, Senior Microbiologist at Yorkshire Water and a member of the Standing Committee of Analysts Working Group 2 on microbiological methods, currently heavily involved with the revision of DoE Report 71, The Bacteriological Examination of Drinking Water Supplies 1982. Shona Cameron, the virologist in Mr. Watkins' team, assisted him in the writing of this paper.

The remaining contributions were of a more specific and practical nature. Jean-Pierre Duguet, Head of Drinking Water Research at Lyonnaise des Eaux, gave a presentation on the French view and experience with novel water treatment methods. Dr. Catherine Benton, who is Principal Microbiologist with Strathclyde Regional Council, was well placed to speak on practical lessons learned from water borne outbreaks of cryptosporidiosis, as she was the water industry microbiologist on the expert group chaired by Sir John Badenoch, set up to advise the government on the significance of cryptosporidium in water supplies. Dr. Benton is presently a member of the group directing the DoE Cryptosporidium Research Programme and is also actively involved in the revision of DoE Report 71.

Much of the UK research into the occurrence and significance of blue-green algae and their toxins is being carried out at the University of Dundee by a team headed by Professor Geoff Codd. Linda Lawton, a member of that team, presented a paper jointly written with Professor Codd on this subject.

The final paper presented by David Foster and Tony Rachwal of Group Research and Development and Stephen White of the Water and Environmental Science section in Thames Water plc, gave a thorough review of new treatment methods being considered and applied for control of pesticides and chlorinated organic micropollutants in drinking water. They are well positioned for this, being closely involved in the development of Thames Water's plans to spend in excess of £200 million on advanced water treatment during this decade.

I believe that readers from all disciplines involved in the future development of potable water treatment will be interested in this book which consists of the papers presented at the symposium and aims to bring together in one concise volume a valuable collection of state of the science contributions on issues of real importance which link drinking water quality and public health.

My thanks are due to the committee of the Tyne and Humber branch of the Institution and particularly to the then chairman, Tom Rae, for encouragement, support and assistance before and during the symposium and subsequently during the process of bringing this book to fruition. Also to the authors and presenters without whom the whole project would have been impossible.

The symposium was generously sponsored by Water Training *international*. In addition to financial support they kindly allowed me to devote a considerable amount of time throughout the last year to the symposium organisation and the subsequent process of editing this book.

Finally, my especial thanks must go to my secretary, Wendy Connelly who took a large part of the burden of the administration of the symposium and has single handedly produced all the text for this book. It is to her considerable credit that, despite all the pressures and frustrations of combining this extra work with an already busy workload, we have come to the end of the project on time and still on speaking terms!

R. Holdsworth
York, April 1991

1. Water quality - does 'don't know' mean 'safe'?

J.W. Bridges, D. Wheeler and O. Bridges

Concern with the quality of life is now high on the political agenda throughout Europe and it is anticipated that it will increase considerably in the 1990s. This is particularly expressed in the UK by public anxiety about contamination of the vital constituents of life: food, air and water. A reflection of this, as far as water is concerned, is the enormous increase in the sales of bottled water, from 3 million per annum in 1976 to 120 million in 1986. In addition, nearly 10 per cent of the population use water filters in an attempt to purify their tap water. Both markets are still expanding.

A public opinion survey conducted by Robens Institute in 1988[1] confirms the public worries about drinking water. Nearly 44 per cent of those interviewed expressed their concern about health risks from drinking water and 28 per cent stated that they always avoided drinking water straight from the tap. There appears to be similar public concerns about the safety of bathing waters.

Obviously, perceptions of risk may differ greatly from the actual risk. Nevertheless, it is these perceptions that are the principal generators of the pressures for improvements to be made. In this article we will concentrate on objective criteria relating to human health risks arising from the ingestion of and dermal contact with water supplied to domestic premises. The quality of bathing waters and implications of pollution of water sources for environmental species is covered by other recent reports.[2,3,4,5]

Historic aspects

Scientific and public interest in water quality is of course not new. In the UK it probably has its origins in the mid 18th century.[6] In 1828 the editor of Hansard, John Wright, published anonymously a pamphlet attacking drinking water quality in London. This led to the setting up of a Royal Commission which established the principle that water for human consumption should at all times be 'wholesome'. The term 'wholesome' has been incorporated into virtually every piece of legislation concerned with drinking water ever since. The extent to which Wright succeeded in raising public consciousness of water quality is indicated by the remark of a contemporary that 'the whole town was in convulsions under the notion that they should be poisoned by filthy water'.

The first unequivocal demonstration of waterborne transmission of the disease cholera was by Snow in 1854. This stimulated great advances in water treatment practice, in particular the routine application of slow sand filtration and disinfection of public water supplies.

Although the Royal Commission of 1828 was concerned with water quality, it had difficulty in defining it precisely, there being virtually no analytical techniques available at the time with which to determine either microbial or chemical contamination. Consequently, since that time there has been a continuing, often fierce debate on what constitutes a suitable quality for human drinking water. Not surprisingly, in the 19th and early parts of the 20th century the evaluation was largely based on subjective, usually sensory perception.[7]

Many authorities (eg Sir Edwin Chadwick) believed an atmospheric miasma above the water rather than the water itself was responsible for disease transmission. As a consequence, great efforts were made to remove the smell, it being assumed that this would dispel the disease. In 1856 during the 'Great Stink', sheets drenched in chloride of lime were hung from the windows of the Houses of Parliament to exclude the smell. At least it focussed the minds of the politicians on the need to take action to improve the quality of London's water supply! It has been speculated,[6] that the construction of municipal trunk sewers by the Victorians was strongly influenced by the desire to remove the miasma.

Even today taste, smell and appearance (colour, turbidity, etc) are considered useful criteria for judging water quality.[8] However, in addition there are now objective methods for determining the presence and levels of many, but by no means all of the microbial contaminants likely to be present in drinking water.

Since the 1960s the emphasis regarding drinking water quality, in the developing countries, has shifted from the bacteriological quality to identification of chemical contaminants. This reflects largely the very considerable success of the water industry in overcoming bacteriological problems, although this victory is not complete (viz many viruses and cryptosporidia).[9]

With the great improvements in analytical chemistry methodology over the past twenty five years has come the growing realisation that normal drinking water contains trace amounts of several thousand chemicals and that only the limitations of these techniques restricts the number of chemicals identified. Many of these chemicals, of course, are of natural origin but pesticides, human and veterinary drugs, industrial and domestic chemicals, and various products arising from the transport and treatment of water are very commonly found, albeit normally at very low concentration.

A similar scenario is developing for air and food. In both cases the number of chemicals identified continues to increase apace.[10,11] The question must be asked as to whether exposure to these chemicals constitutes any significant risk to human health. This question has to be set in the context that we do not as yet know with certainty the principal causes of the majority of the chronic diseases which beset the UK population. Estimates of the relative contribution of chemical exposure in any form to the nature and incidence of chronic disease varies widely from a few per cent at most to being attributed as the principal cause.

In addressing the problem of the contribution of waterborne chemicals to the incidence of human disease, water scientists, whose previous experience typically has been largely confined to dealing with microbiological problems, have tended to focus on acute risks. The absence of detectable, short term adverse effects of drinking water on human health has been taken by many as conclusive evidence that the presence of such chemicals is without risk to man.

While information on the acute toxicity of a chemical can be very useful in determining the response to an emergency situation[12] such as an accidental spillage or deliberate release of chemicals into the water supply, such information is of very little help in predicting the effects of daily exposure to a chemical over many years. Indeed, as has been shown only too clearly from the aftermath of the Camelford incident[13] the existing data base has even proved inadequate to identify the adverse health effects of short term exposure to high levels of a chemical, aluminium sulphate, which is widely used by the water industry itself.

However, low levels of chemicals are much more likely to cause chronic than acute effects on health. Here direct reliable information is very sparse. Some authorities appear to have accepted the extremely naive assumption that information on the acute effects of a chemical, in either man or in experimental animals, can be used to predict the effects of being exposed to it for a lifetime. This is exemplified by the very recent DoE Consultation paper on classification of substances in ground water.[14] This paper recommends that for a change from list I to list II status acute oral LD_{50} data values are used, an LD_{50} of 500mg/kg in the most susceptible mammal being the determinant. It has to be emphasised that in practice only rarely do the chronic effects of a chemical have any resemblance to the acute effects. Although there is an obvious need to have a regulatory standard and acute data are much more readily available than chronic data, is it reasonable to use one that provides for the non-expert a false state of assurance? Despite its limitations (see below) a sounder approach toxicologically would be to use workplace data.

If we are to obtain a proper assessment of the health risk that could arise in man through exposure to chemicals in water over a lifetime, information must be developed on the following:

i)   identification of chemicals which are of most concern;
ii) data on the effects of long term exposure in man and/or animals to each chemical;
iii) a measure of the extent and form of exposure to each chemical;
iv) identification of particularly at risk groups;
v)   the means of establishing how exposure to other chemicals in the water can modify the toxicity.

Chemicals of particular concern

The chemicals of most concern are likely to be those with one or more of the following properties: high intrinsic chronic toxicity, persistence in the body, bioaccumulation in the food chain, significant magnitude and frequency of possible contamination.

Unfortunately, only a modest percentage of these chemicals present in normal drinking water has been identified. Information on the others may come from a variety of sources. These include:
- epidemiological investigations among the general population;
- assessment of the effects on the workforce in industries which have a particular use of one or more chemicals;
- long term studies of a chemical in animal models;
- data on chemicals with very closely related physico-chemical properties.

Epidemiological investigation

These have become a very important aspect of identifying the effects of various substances, eg tobacco smoke, on human health. However, as a technique it has a number of limitations which restrict its value for identifying possible adverse effects on human health from water.

One is the need for a prior hypothesis. This hypothesis must be carefully framed, the endpoints for assessment set and the measurement of the particular chemical or group of chemicals of concern established. Only a few, preferably one hypothesis can reasonably be examined at a time. If a number of possible outcomes are defined, each strictly represents a separate hypothesis. At the extreme, if there are 20 possible effects of exposure examined which were independent of one another then as a result of chance alone (at $p < 0.05$) one will show an association with the exposure.[15] It is only too tempting to examine one health effect or one chemical after another until 'something turns up'. It is important that if something is discovered in this way the newly 'generated' hypothesis will need to be tested in a further study.

Another limitation is the selection of a representative population. In the case of direct investigations on water this requires the identification of discrete high and low exposure groups. Having selected these populations a substantial number of randomly selected individuals is required from whom to choose the study participants.

Other important issues include the selection of the methodology for ascertaining the effects on health and the actual levels of exposure. In addition, it is important to identify other possible sources of exposure to the chemical or chemical group. This is a most difficult task since few if any chemicals are confined solely to water.

Because of the above difficulties it is unlikely that epidemiological studies on their own will be in a position to answer the question directly as to whether or not the range and levels of chemicals currently found in water causes any long term adverse effects on human health.

Chemicals in the workplace

Use of information derived from investigations (including epidemiological investigations) of the effects of particular chemicals on the health of those working in particular industries has proved a valuable data source.

Some data at least are available through this means on the long term effects of several hundred chemicals. Indeed this data is used to set safe levels for airborne exposure to workplace chemicals.[16] However, it has to be appreciated that the findings can be applied directly to a fairly restricted range of the population. The exposed group does not include the young, the old, and those with many types of disease. Indeed, the group may be confined to a single sex and social class too. In addition, inhalation is generally the route of exposure investigated and the actual levels of exposure are often difficult to identify. Nonetheless, findings from workplace studies have highlighted the toxic risks from many chemicals which may be found in the water supply, including arsenic, cadmium, chromium, the chlorinated aliphatic hydrocarbons, benzene, polycyclic aromatic hydrocarbons and asbestos.

Animal studies

Undoubtedly, the main source of information on the possible toxicological effects of chemicals is from investigations in animal models, typically rats, mice and dogs. For the foreseeable future this is almost certain, for most chemicals, to be the only available form of information on their chronic toxic properties. As with the other sources of data referred to above, animal studies have their particular limitations as far as predicting the adverse effects of trace amounts of chemicals in water is concerned. Firstly, animal toxicity tests are normally conducted at relatively high doses. Consequently, the relevance of the findings to the much lower levels which water consumers might be exposed to is often very questionable. Secondly, animals and man may respond differently to a number of chemicals, hence extrapolation from animal toxicological findings to the human situation may be inaccurate. One aspect of this is that animal studies do not tend to pick up allergic and sensitisation effects, or the subtle central nervous system effects which certain chemicals are known to produce in man.

For risk assessment purposes chemicals are commonly placed into one of two categories. These are those for which a threshhold exposure level for toxicity can be established and those where no safe exposure level can be ascertained. The latter appears to be the case for many genotoxic carcinogens and may also be true for chemicals which produce true allergic sensitisation. The approach to identifying an acceptable level is different for the two categories. For the former a standard can be set based on the threshhold dose in man (or alternatively in the most sensitive animal model). Normally an additional safety factor is built into the standard to allow for the facts that some individuals are likely to be more sensitive to the effects of the chemical than others, and/or the studies being relied on are either not directly applicable (eg the exposure situation is different or inter-species variations are likely) or not definitive. For chemicals for which no safe exposure level can be ascertained it is necessary to identify a 'virtually safe' exposure level instead. In the case of cancer this is often defined as the dose which will on exposure for a lifetime, increase the probability of cancer occurring by 1 in 100,000 or 1 in 1,000,000. In the United States quantitative risk assessment models are favoured as a means of deriving this virtually safe dose.[17] In Europe in general, and in the UK in particular, a rather more 'rule of thumb' approach has remained in being, but this is likely to change!

The data base for identifying bioaccumulation in the food chain and persistence in the environment is, unfortunately, just as deficient as that for assessing the chronic toxicity of chemicals. Even where biodegradation studies have been conducted, often there has been no adequate characterisation of the nature of the products formed and their possible effects on health.

Identification of particularly at risk groups of the population

The very young, the elderly, those with severe disease and pregnant women ought to be considered generally as possibly at risk groups.

As indicated above, identification of at risk groups for a particular chemical is an aspect of determining the size of the safety factor to be used in setting a standard. Unfortunately there is very little reliable data on the magnitude of the increased sensitivity of at risk groups to even common chemicals. In the absence of such data the widely adopted regulatory approach has been to assume that it is embraced within a ten fold range and therefore a safety factor of 10 is used to allow for inter-individual variations in response.

Exposure to mixtures of chemicals

One of the particular dilemmas in setting a standard is what allowance should be made for the presence of other chemicals which might also exert some toxic effects. In principle, chemicals in combination may produce additive effects or one might synergise or inhibit the action of another. Although there has been much discussion on this issue for many years, there has been remarkably little action by legislative authorities to do something about it in terms of either promoting research or the rational application of first principles to chemical mixtures. For workplace chemicals the Health and Safety Executive has suggested[16] that, in the absence of data to the contrary, chemicals should be assumed to be additive in their effects. If this logical approach was adopted for water, as indeed was proposed in America by the Committee on Water Quality Criteria of the National Academy of Sciences as long ago as 1972, the following equation would be applied in ascertaining the overall health risk from the chemicals present in water:

$$\frac{a}{S_a} + \frac{b}{S_b} + \frac{c}{S_c} \ldots\ldots \frac{n}{S_n} = \frac{x}{S_x}$$

Where a, b, c, etc. are the concentrations of individual chemicals found in the water and S_a, S_b, S_c etc are the safety standards set for each. If x/S_x exceeds unity then the water quality would be questionable and steps would be required to reduce the levels of one or more components. The EC drinking water directive[18] has for the first time as far as water is concerned, introduced the principle of regulating mixtures of chemicals as well as individual chemicals by setting a standard for total pesticides as well as individual pesticides. Although it has to be said that the approach adopted is simple in principle, and useful from a regulatory viewpoint, in toxicological terms it is nonsensical as well as being illogical from an analytical standpoint.

The emphasis of the above discussion has been on the risk from ingestion of water. However, exposure to chemical contaminants in water may also occur through dermal and inhalation exposure. For inhalation exposure the main concern is the volatile components such as chlorinated aliphatic hydrocarbons which may be released, for example, when taking a shower.

It is apparent from the above discussion that our current level of understanding is seriously deficient in regard to possible adverse effects on human health which may result from consuming drinking water for a lifetime. Considerations of whether or not it is safe therefore depend on which of two standpoints is adopted, either that in the absence of data to the contrary it should be viewed as safe, or because the data is lacking it should be considered suspect (the so-called precautionary approach).

Either way it is very clear that much further research is needed to confirm the absence of long term adverse health effects. Bearing in mind the essential role of water for human life, the current research effort to remedy this situation is remarkably low.

In a number of cases the necessary work has, of course, already been done but the remedial action has been slow. This is undoubtedly true of the replacement of lead pipes in soft water areas. In such situations some justification of the delayed action in terms of the risk/benefit assessment used in comparison with other remedial projects should be available for discussion.

For a number of other chemicals, such as nitrate, polycyclic aromatic hydrocarbons, and aluminium, there is considerable public concern but insufficient scientific data to make a judgement on priorities. The issues are briefly summarised below.

Nitrate

For nitrate the principal concern is whether high levels of nitrate in drinking water have a significant influence on the rate of formation of carcinogenic nitrosamines in the stomach and small intestine. To date, the epidemiological investigations have concentrated largely on whether or not there is a significant increase in stomach and/or other intestinal cancers, with conflicting results. However, if increased amounts of nitrosamines were formed, animal studies indicate that the gastrointestinal tract will not necessarily be the primary target, nor will cancer be the sole adverse effect.

Polycyclic aromatic hydrocarbons

Degradation of coal tar pitch pipe linings can lead to the release of particulate matter into the drinking water. These particles contain high levels of a range of substances including well established polycyclic aromatic hydrocarbons. Examples are benzo (a) pyrene and benzo (b) fluoranthene plus a number of substances with unknown biological properties.[19] In the absence of information to the contrary, it must be assumed that some members of the public regularly could consume water containing rather high levels of coal tar pitch and consequently high levels of various carcinogens. There is some debate on whether these carcinogens would be readily released from the particle in the intestine. However, if a particle became lodged in the gut it could provide a 'hot spot' for the slow regular release of components, a scenario which would encourage cancer initiation. It is not an easy matter experimentally to

Aluminium

Until comparatively recently aluminium was considered to be innocuous.[20] However, the recently postulated apparent association between aluminium levels in the brain with Alzheimer's disease and the distressing aftermath of the Camelford incident, has thrown open the question of the magnitude of any risk from aluminium exposure in drinking water. Also still in question is the importance of drinking water as a principal source of aluminium exposure. Meanwhile doubts have also arisen about the safety in use of other flocculant coagulants based on polyacrylamide because of their acrylamide monomer content.

Other chemicals

Contamination of ground water has become a major public issue in the United States and threatens to become so in the UK. Major sources of contamination are agricultural and industrial activities and leachates from toxic waste disposal sites. The latter source is of particular concern because the location of many of the sites, of which there may be several thousand, is in many cases uncertain and there is no record of which chemicals may be deposited there. Consequently, the potential risk from the contamination of water is unknown and the development of monitoring procedures to gauge this is fraught with problems. The degree of uncertainty about the toxic hazard from industrial effluents is not properly recognised. In the majority of cases the composition of such effluents is not fully characterised and the toxicological properties of the known components are poorly evaluated. For various reasons new ways of dealing with such industrial wastes are likely to be required. Firstly, because there will be growing pressure on industry from the public and the legislators to limit waste production. Secondly, because it will become increasingly unacceptable in many areas to dispose of industrial waste along with domestic sewage, because such a strategy limits the use for agricultural and other purposes of the sludge produced as a byproduct of sewage treatment.[21]

In the agricultural arena an aspect of growing concern is the use of potent antimicrobial agents on fish farms located in close proximity to sources of drinking water. The issue here is the extent of the dangers engendered by the regular flushing of these chemicals from such farms. Inappropriate disposal of pesticide residues, for example used sheep dip liquors by farmers, is also coming increasingly under the microscope.

Future prospects

In the absence of suitable biological systems which can live in the water supply without contaminating it and which can serve as sensitive indicators of the chronic toxic effects of the water constituent, it is inevitable that increasingly comprehensive monitoring will be demanded. In the foreseeable future simple water test kits for specific chemicals are likely to become available to the public and the use of these will place strong additional pressures on the water companies.

An evaluation of all health risks associated with drinking water is timely, with initial attention being focused on each chemical contributed by the water industry itself as a consequence of water transport, treatment and storage.

The strategies for dealing with emergency incidents also need to be carefully re-evaluated and regularly updated. These emergency plans should give particular attention to communication systems, monitoring strategies, and rapid identification of at risk members of the public.

References

1. Robens Institute (1988). *Public Attitudes to Water and Water Privatisation.* University of Surrey, UK.
2. United Nations (1990). *Global Outlook 2000.* United Nations, New York.
3. Meybeck, M., Chapman, D. and Helmer, R. (1989) *Global Freshwater Quality. A First Assessment.* Blackwell, Oxford, UK.
4. Brown, J.M., Campbell, E.A. Rickards, A.D. and Wheeler D, (1987). 'Sewage Pollution of Bathing Water'. *Lancet* ii, pp 1208-1209.
5. Wheeler, D. (1990). 'On the Beach'. *Laboratory Practice*, 39, 4, pp19-24.
6. Wheeler, D. (1989). PhD Thesis, University of Surrey.
7. Wheeler, D (1986). 'Britain's Polluted Drinking Water'. Report. *The Ecologist* 16, pp130-131.
8. WHO, *Guidelines for Drinking Water Quality* (1984-1985) Vols.1-3 WHO Geneva.
9. Department of the Environment and Department of Health. *Cryptosporidium in water supplies*, HMSO, London, July 1990.

10. Grant, L., Duk Lee, S., Schneider, T., Wolters, G.J.R., and Mehlman, M.A. Ed. (1990). *Health and Environmental Hazards from Toxic Chemicals in Air*, Princeton Scientific Publishing Co. New Jersey, USA.
11. MAFF, *Food Surveillance Reports*, HMSO London.
12. Bridges, J.W. (1990). Identification of Toxic Hazard in Major Chemical Disasters - Medical Aspects of Management, Ed. Murray V, RSM London.
13. Anon (1988). Water Poisoning in Cornwall ii 465.
14. *Consultation Paper on Proposed National Classification Scheme of Listed Substances for Purpose of EC Groundwater Directive*, (1990). DoE, October.
15. Elwood, P.C., (1986) Interpretation of Epidemiological Data-Pitfalls and Abuses in Toxic Hazard Assessment of Chemicals, Ed. M.L. Richardson, Royal Society of Chemistry.
16. Health and Safety Executive Guidance Note EH40/90.
17. Johannsen, F.R. (1990) 'Risk assessment of Carcinogenic and Non Carcinogenic chemicals'. *Toxicology* 20, pp 341-367.
18. *Council of European Communities Directive relating to the quality of water intended for human consumption* (80/778/EEC) July 1980.
19. Crane, R.I., Fielding, M., Gibson, T.M. and Steel, C.P. (1981). *A Survey of Polycyclic Aromatic Hydrocarbons in British Waters*, WRc Report.
20. Birchall, J.D. and Chappell, J.S. (1988) 'Aluminium, Chemical Physiology and Alzheimers Disease'. *Lancet*, October 29th, pp 1008-1010.
21. Wheeler, D. and Merrett-Jones, M. (1991). Sewage Sludge - Resource of the Future?, *Greenpeace*, London.

2. Developments in health related quality standards for chemicals in drinking water

J. K. Fawell

Introduction

One of the major constraints in the process of supplying drinking water is that of water quality standards and in some cases the lack of them. Relatively small differences in standards or their application can make very great differences in the costs and problems of water treatment or supply in particular circumstances. One example of this is the nitrate standard of 50mg/l nitrate[1] and the recent change of interpretation in the UK from an average figure over a period of time, to take account of variations in concentrations, to the requirement to achieve 50mg/l as an absolute maximum.

A further problem facing water suppliers and regulatory authorities is that of what action to take when a contaminant is present at concentrations above the water quality standard. Should the supply be terminated with the consequent hazards from losing the supply altogether? Or should suppliers consider the possibility of introducing water of potentially lower quality, by changing to sources which have not been used for some time. This may bring quality problems associated with, for example, re-commissioning mothballed pipework or treatment plant? Such a situation also leads to the difficulty of reassuring consumers who are constantly told by the media and pressure groups that any breach of any standard is a threat to health. There are numerous examples of such situations, including failure to meet standards based on political considerations such as those for pesticides and particularly the standards based on aesthetic considerations.

However, water suppliers are also faced with the problem of contaminants for which there are no standards. In such cases there are problems in determining whether a water is suitable for continued supply or whether some actions should be taken. Even more acute is the problem of reassuring consumers in a climate in which water is no longer universally regarded as being safe and in which the water suppliers are regarded as profit making corporations, not benevolent public services.

Approaches to setting standards

One of the main sources of confusion regarding drinking water standards and their interpretation is the lack of any clear indication as to how the standard was derived. This results in interpretation of all standards as 'health' standards by the public and in difficulty in assessing what should be done by the water supplier if a standard is exceeded. This is particularly true of the EC directive on drinking water[1] for which the derivation of the actual numbers is given no explanation. There are even numbers for parameters regarded as toxic which are based on political or other considerations only loosely based in science. The two parameters concerned are pesticides, parameter 55 in the directive, and polycyclic aromatic hydrocarbons (PAH), parameter 56. The use of such approaches is perfectly acceptable as long as the reasoning behind them is clear to all.

Currently the World Health Organisation (WHO) Guidelines for Drinking Water Quality[2] are being revised. These guidelines are intended to enable governments to use them as a basis for standards, taking into account local conditions. They are intended to be protective of public health and they will be presented in such a way that all of the stages will be absolutely clear, even down to the detailed scientific considerations such as derivation of uncertainty factors and the rounding of numbers. It is therefore incumbent on the expert groups to justify fully their thinking and present it openly for all to see. Such a discipline avoids the fudging of issues whilst giving the impression of scientific precision and can only be of value in increasing public confidence in the resulting guidelines.

The scientific basis of standards

In order to derive health based standards it is necessary for there to be adequate data on the effects of the chemical concerned on humans or laboratory animals.

What constitutes adequate data will depend on the particular circumstances, for example there is a substantial database on the effects of aluminium based food additives in laboratory animals. Indeed the FAO/WHO Joint Committee on Food Additives and Contaminants have recently proposed a provisional tolerable weekly intake (PTWI) for aluminium of 7 mg/kg body weight.[3] However, concern over waterborne aluminium relates to compounds which are chemically very different and which may present a different profile in terms of bio-availability. In addition, the question of whether aluminium from any source is causally related to Alzheimer's disease has still not been resolved. This is of particular importance with regard to water. Allocating a proportion of the PTWI to water will give a figure in excess of that needed to protect against aesthetic effects and may be misleading if intended to protect public health. At present it would therefore not be possible to set a standard, soundly based on health related data, for aluminium in potable water.

Assuming that sufficient data of appropriate quality are available, it then becomes necessary to apply these to the derivation of a standard. At first sight it would seem most desirable to use human data but, although supporting human data are extremely useful, there are difficulties in using them to derive standards. In the case of epidemiological studies, occupational exposure frequently results from exposure by a different route to the consumption of drinking water, for example inhalation or absorption through skin and environmental epidemiological studies usually suffer from very poor measurement of exposure concentrations. With clinical studies it is not usually possible to determine a true 'no effect level' since it is clearly unethical to give a sufficiently high dose to obtain clear effects. Such a situation has arisen with chlorate and chlorite where the National Academy of Sciences has attempted to use data from volunteer studies, but the top dose used in the study gives no indication of how high the margin of safety is because it does not produce any measurable effects. In addition, the study is very short so that effects resulting from longterm exposure may not be apparent.

We then have to rely on data from studies on laboratory animals with all the associated problems. However, in order to use the results of laboratory animal studies in toxicology, it is necessary to use a means of extrapolation to derive a safe level for lifetime exposure of a human population.

There are two available techniques. One is the application of safety factors, preferably called uncertainty factors, to what is usually a no observed adverse effect dose. This approach is applicable for organ-specific toxicity, including reproductive and behavioural toxicity, for which it is generally believed that there is a dose below which no adverse effects will occur. The second technique involves the application of mathematical models for low dose extrapolation. These are only used in the case of certain carcinogens.

The initiating event in the process of chemical carcinogenicity is usually considered to be the induction of a mutation in the DNA of a somatic cell. In some cases the chemical acts directly on the DNA, in other cases it may require activation by a metabolic system. The process theoretically may not have a threshold, or at least not one which can be practically ascertained. The United States Environmental Protection Agency (USEPA) and WHO use the linearized multistage model for low-dose extrapolation, although there are a number of other models. The models give a maximum likelihood estimate (MLE) and a 95 per cent confidence limit at different levels of risk. The USEPA use this latter figure for promulgating standards. However, this is the upperbound value and is unlikely to underestimate the true risk, which could be zero at the other extreme if there is a practical threshold to the toxic mechanism.

There are however, some substances which are capable of producing cancer in animal studies but which cannot be shown to have genotoxic activity. With such non-genotoxic carcinogens, there is believed to be a threshold dose and the use of low-dose extrapolation models would therefore not be appropriate. This concept has been accepted by WHO in the revision of the guidelines and they have given clear guidance to the countries preparing the support documents. Indeed the USEPA also seem to have accepted this concept with their new standard for p-dichlorobenzene[4], for which there was clear evidence of carcinogenicity in rat kidney and mouse liver. However, the compound is non-genotoxic and it has been regulated using an uncertainty factor approach, with an additional uncertainty factor for the carcinogenic endpoint. The resultant figure is however well above the taste and odour threshold.

Additional assumptions in setting standards

There are a number of considerations which may make a substantial difference to the final figure to be used for regulatory purposes. One of these is the size of the person assumed to be protected.

Acceptable daily intakes (ADI) are calculated on the basis of intake per day and therefore this needs to be translated to a 'standard' individual. The USEPA and WHO assume a 70kg adult, whilst in the UK we more frequently use a 60kg adult. However, there may be circumstances when children are a high risk group and since they have a higher intake of fluid per kg body weight, then child weight may need to be the assumption. This is usually taken as 10kg.

The assumption for water intake is usually 2 litres per day for an adult but 1 litre per day for a child. In temperate developed countries this is a figure which would encompass the majority of the population. However, in hotter countries this may need to be adjusted and standards of course are intended to protect 'normal' use of water and not special groups, such as dialysis patients, who may require special standards and point of use treatment to provide water of adequate quality. Aluminium is of course a good example of a contaminant requiring a more stringent standard for dialysis, with a figure of between 10 and 30 ug/l being desirable.

One of the problems, which has come to the fore with drinking water standards for pesticides, is that of making allowance for other sources of exposure, such as food. WHO have used an arbitrary allowance of 1% of an ADI for compounds to which significant exposure by food might be expected, but 10% for those to which exposure from food would be less.

Consequently this assumption can make a difference of an order of magnitude to the proposed guideline or standard. WHO are therefore attempting to obtain as much information as possible on levels of pesticides in food.

The allocation of a proportion of an ADI to food for pesticides however, immediately begs the question, 'what about other chemicals?' USEPA have tackled this by allocating 20 per cent of an ADI or TDI (tolerable daily intake) to water for chemicals other than pesticides. Such an approach makes some allowance for occupational exposure and exposure from food. This question has still not been wholly resolved but will be taken into account by WHO. The problem of whether to allocate a proportion of the 'standard' for carcinogens to other sources has been addressed, but at present the conservatism of the low-dose extrapolation models is considered to be sufficiently protective of public health.

The final consideration is the contribution from water to exposure by inhalation and dermal absorption. In most cases this does not appear to be sufficiently great to warrant additional allowance in the derivation of standards, however for some parameters such as radon degassing from water will be a major route of exposure.

Some examples of uncertainties in setting standards

The above discussion can best be illustrated by an examination of some contaminants for which revision of standards is under consideration by WHO and in other countries.

Lead

Data on the neurological effects in children of exposure to low levels of lead has continued to accumulate and indicates that such effects are detectable at lower concentrations than previously supposed. WHO have examined the evidence and have proposed a PTWI of 25 ug/kg body weight in children. However, there are also proposals to limit population exposure by assessing blood lead levels. The difficulty with the latter approach is the problem of assessing the relationship between lead intake, lead absorption and the concentration in the blood, since this relationship is complex and may be curvilinear. However, the difficulty with both approaches for the regulator concerned with drinking water is how to allow for exposure from other sources than water.

Chloroform

Chloroform produces liver and kidney tumours in mice, and kidney tumours in rats. Although these effects are quite clear, it is very questionable as to whether chloroform is genotoxic and whether these findings are appropriate for low-dose extrapolation using mathematical models. There are data which indicate that liver and kidney tumours produced by chloroform (in mice) are a consequence of repeated tissue damage. This results in cell division to replace dead cells, a process which is more susceptible to naturally occurring mutations. The induction of kidney tumours in male rats is sometimes associated with blocking of the degradation of a protein, alpha-2u-globulin, which concentrates in the kidney and causes cell death with subsequent reparative cell division.

Such detail may seem unnecessary, but the potential impact on regulatory levels can be startling as shown below. It would therefore be worthwhile carrying out the research required to discover the most appropriate means of determining a safe level for regulatory purposes.

Table 1   Chloroform

Level of Risk	MLE	Upper 95 per cent Confidence Limit
10^{-6}	104 ug/l	9 ug/l
10^{-4}	10,400 ug/l	900 ug/l

NOTE: Uncertainty factor of 10,000 to lowest dose giving kidney tumours in male rats gives a figure of 480 ug/l.

Other trihalomethanes

Bromodichloromethane is similar to chloroform except that it also produces intestinal tumours in male and female rats which are normally relatively rare in the strain of rats tested. In addition, the rat kidney tumours do not appear to be associated with the accumulation of alpha-2u-globulin. The evidence for mutagenicity is, however, weak and when the MLE is compared to the 95% confidence limit in the linear multistage model for low-dose extrapolation, they are substantially different, indicating a poor fit to the data.

Chlorodibromomethane induces only liver tumours in mice and there is only equivocal data on mutagenicity. The mechanism of carcinogenicity is suspect because the strain of mouse used has a high spontaneous incidence of liver tumours and the doses used in the carcinogenicity study all produced liver damage. Bromoform induced a low incidence of intestinal tumours in rats but these tumours, as mentioned above, are relatively rare. The induction of tumours in only one species is not strong evidence of carcinogenicity and there are indications that the data fit to the linear multistage model is poor. It is therefore possible that using this approach for regulation would be inappropriate.

The risk levels given by the linear multistage model for the other THMs are given below in Table 2.

The position is similar with a number of other chlorinated compounds such as trichlorethylene and tetrachlorethylene, for which there is evidence of a non-genotoxic mechanism of carcinogenicity. This, if it were supported by adequate mechanistic data, could lead to an appropriate regulatory level considerably above that derived using the linear multistage model.

Table 2   *Other trihalomethanes*

	Level of Risk	MLE	Upper 95 per cent Confidence Limit	10,000 Uncertainty Factor
Bromodichloromethane	10^{-6}	10 ug/l	0.5 ug/l	54 ug/l
	10^{-4}	1,000 ug/l	50 ug/l	
Chlorodibromomethane	10^{-6}	20 ug/l	0.42 ug/l	54 ug/l
	10^{-4}	2,000 ug/l	42 ug/l	
Bromoform	10^{-6}	81 ug/l	5 ug/l	108 ug/l
	10^{-4}	8,100 ug/l	500 ug/l	

Discussion

Determination of standards and guidelines for drinking water based on toxicological data has progressed as knowledge of the basic science and understanding of toxicology has advanced. This may lead to changes in some of the existing standards, but only if there is considered to be sufficient data to support such a change. However, it is most important that the scientific thinking behind standards and particularly behind any change, is made absolutely clear and that the assumptions incorporated in the calculations are not hidden.

Unfortunately this means that time and effort must be expended in carrying out the detailed mechanistic studies and of course this may also result in some older standards being lowered because they may not be considered to be adequately protective of public health. Openness means that the gaps in the data must be admitted and the risk is that the public will interpret such gaps as indicating a serious problem, taking the view that no news is bad news.

However, in the final analysis, the benefits of health based standards, underpinned by good science and a completely open approach will heavily outweigh the short term disadvantages.

References

1. *Council of European Communities Directive relating to the quality of Water intended for human consumption* (80/778/EEC) July 1980.
2. *Guidelines for Drinking Water Quality,* ,World Health Organization, Geneva, (1984).
3. FAO/WHO. The evaluation of certain food additives and contaminants, 33rd report of the joint FAO/WHO Expert Committee on Food Additives, 1989. Technical Report Series 76, pp 26-27.
4. USEPA Water Pollution Control; National Primary Drinking Water Regulations - Volatile Synthetic Organic Chemicals; Para-dichlorobenzene Federal Register, Vol. 52, No. 74, pp12876 - 12883 1987.
5. FAO/WHO 'Toxicological Evaluation of Certain Food Additives and Contaminants', 30th meeting of joint FAO/WHO Expert Committee on Food Additives, Rome, 2-11 June 1986.

3. Recently recognised concerns in drinking water microbiology

J. Watkins and S. Cameron

Introduction

Water quality is of paramount importance to protect public health and no aspect of that quality is more important than the microbiology. The treatment of drinking water and the separation of drinking water and sewerage systems has been of prime importance in the reduction of waterborne disease in developed countries. However, waterborne and water related diseases still occur, if infrequently, and their occurrence is recognised and vigilance in water treatment and water quality monitoring has been stressed.[1,2] There are questions about our ability to assess the presence of waterborne pathogens using current bacteriological indicators, the time lapse between sampling and the availability of the result and the ability to recover environmentally damaged or treatment damaged microorganisms. There are newly emerging pathogens which require ever more difficult and demanding techniques for isolation. Water quality may deteriorate in the distribution system due to ingress and contamination, or the growth of organisms which can take advantage of elevated temperatures and organic nutrients.

There is need for a better understanding of the incidence, survival and distribution of microorganisms in the water environment, for the protection of users of surface waters for recreational purposes, and of the effects that water treatment has on those organisms, for the protection of the consumer. Treated water is also used in a variety of ways and such use can adversely affect water quality.

Waterborne disease

Outbreaks of waterborne disease have been well documented. Thirty-four outbreaks were identified in the United Kingdom between 1937 and 1986 excluding the typhoid outbreak at Croydon.[3] Twenty one of these outbreaks occurred in public supplies affecting nearly ten thousand people. A similar review for Scotland covering the period between 1945 and 1987 identified fifty-seven outbreaks.[4] Eighteen of these occurred in public supplies affecting five thousand five hundred people. Deficiencies which gave rise to outbreaks were identified as untreated or inadequately treated surface water, contamination in distribution and contamination due to cross connections or mains repair.

An outbreak of *Cryptosporidiosis* in the United States affected an estimated thirteen thousand people.[5] Water quality was within the limits for coliform bacteria, chlorine levels and turbidity. An outbreak of Rotavirus infection in China in 1982-83 affected more than twelve thousand adults.[6] The water supply in the area was found to be bacteriologically contaminated and the pattern of spread suggested water as the original source of the outbreak. Enterovirus and Hepatitis A virus caused an outbreak in the United States in 1980 with an attack rate of seventy nine percent in a ten thousand population.[7] The outbreak was associated with an underground water supply receiving only chlorination. Viruses were isolated from the supply in the absence of faecal indicators and with 0.8mg/l of free chlorine.

There are many reasons why outbreaks can occur. Many microorganisms have a lower susceptibility to chlorine than faecal indicators, in particular the viruses and the intestinal parasites. Reliance upon indicator testing, particularly where disinfection is used to meet water quality standards, may give a false sense of security. Some microorganisms can be protected from the action of chlorine by organic material, slimes or even association with free-living amoebae. Sudden unexpected contamination of a water source, pre-treatment contamination, may render an effective treatment ineffective. The spillage of cattle slurry could introduce large numbers of pathogens into a surface water. Water treatment may be ineffective, may suffer from breakdown or may consist solely of chlorination. In the United States where giardiasis is the most common waterborne disease, inadequate treatment and treatment deficiencies are recognised as major contributory factors.[8] The integrity of the distribution system may be in question with possible ingress occurring through service storage reservoirs, fractures or fittings on water mains, burst water mains, cross connections with contaminated supplies or the creation of back syphonage through negative pressure, all examples of post-treatment contamination. The spreading of slurry has been shown to have caused contamination of a drinking water supply resulting in a small outbreak of *Cryptosporidiosis* in Scotland in 1988.[9]

Many pathogens are difficult to detect and the methodology is low in recovery efficiency, time consuming, expensive and not always amenable to routine monitoring. For some, methods are not yet available. Little is known both of the incidence and survival of many pathogens in the environment, and of the effect of sewage discharge and industrial and agricultural practices on the microbiological quality of surface waters. Many pathogens have a low infective dose which may require the analysis of large volumes of water for detection. Additionally, low numbers of organisms in large volumes of water make the demonstration of an epidemiologically significant increase in infections almost impossible. Where water is heavily contaminated however, there may be rapid spread through a susceptible population resulting in several thousand cases in a community outbreak.

The faecal indicator

Faecal contamination of water is traditionally detected by the selective culture of indicator bacteria on media which demonstrate the fermentation of lactose.[10] Such culture requires the indicator to be present in excess of pathogens and it is therefore by inference only that water is deemed to be safe for human consumption in the absence of the indicator. Cultural techniques have their own problems. The viability of the indicator must be maintained between the sampling point and the laboratory, although the bacteria may be in a viable state but not recoverable by culture. The specificity of detection may be imperfect and add to the problem with non-indicator bacteria being able to ferment lactose and indicator bacteria failing to do so.

Possession of the enzyme β-galactosidase was seen as a suitable alternative to the fermentation of lactose,[11] overcoming the problems of anaerogenesis in coliform bacteria. This provides for the use of alternative substrates such as ortho-nitrophenyl - β - D - galactopyranoside (ONGP) which is split with the production of the chromogen ortho-nitrophenol. The enzyme is found in over 98 per cent of coliforms and *Escherichia coli*. The enzyme β-glucuronidase is found in 95 per cent of *E.coli* strains, approximately 40 per cent of *Shigella* spp. and occasionally in streptococci. A similar chromogenic substrate para-nitrophenol - β - D - glucuronide can be used for detection. Fluorogenic substrates are also available, methylumbelliferone - β - D - galactoside for coliforms and methylumbelliferone - β - D - glucuronide (MUG) for *E.coli*. Breakdown of the substrate gives fluorescence under ultra-violet light at 366nm wavelength.

Such substrates have provided an alternative for the confirmation of presumptive coliforms and *E.coli* isolated by membrane filtration. Combination of ONGP and MUG gives a single medium for the specific detection of both. Trials with such media in the United States (Autoanalysis Colilert - AC) have given equivalent results to multiple tube fermentation for coliforms.[12] A further comparison of membrane filtration against AC[13] found good correlation between the two tests for total coliforms but not for *E.coli*. A colorimetric test has been described based upon β - glucuronidase activity which makes it amenable to automated analysis.[14] The method is, however, labour intensive but more rapid than current techniques. A seven hour membrane filtration test has been described using conventional media[15] and a six hour test using 4 methylumbelliferone - β - D - galactoside.[16] Both methods have the limit of detection at one coliform per 100ml. Both methods have promise where results are required urgently, providing suitable provision is made should remedial action be required. The ultimate will be gene probes based upon detecting the DNA sequence which codes for specific gene products. It would not detect viable from non-viable organisms but would serve as a rapid screening test.

Microbial pathogens

The range of microbial pathogens is large,[17] consisting of bacteria, viruses and intestinal parasites.

The bacteria

The bacteria can be divided into two groups. There are those which on entry into the aquatic environment loose their viability or remain viable but cannot be recovered by standard cultural techniques. Examples of this group include *Salmonella* spp., *Campylobacter* spp. and *E.coli*. A second group are able to multiply in the environment if suitable conditions exist. Aeromonads can grow in warm waters with low levels of organic nutrients. Pseudomonads and *Legionella* spp. provide additional examples. They are able to multiply to levels where they become problematic in man-made environments. How many man-made environments are there, or are we likely to create, in which these and other microorganisms can fluorish and cause problems?

The cultural problem The isolation of pathogenic bacteria by culture brings with it a number of problems. Refining techniques to maximise recovery and minimise overgrowth with contaminants often results in a lengthy procedure with poor recovery rates and considerable statistical variance. The situation is complicated by the fact that some bacteria, on leaving the gastro-intestinal system, appear to die rapidly. Research has shown[18] that such bacteria which are starved of nutrient appear to loose viability when monitored by a standard cultural technique but, by the use of a direct viable count, show no decline in numbers. Animal passage has shown that such cells retain pathogenicity and can be recovered by culture from an animal system. This research has been used to show that *Vibrio cholerae* O1 can be found in environmental surface waters in a state which is viable but non-culturable.[19] Counts were much higher than could be obtained by cultural techniques. Similar results were obtained in studying the association of *V.cholerae* with aquatic plankton[20] and detecting viable but non-culturable *Salmonella* spp. in chlorinated wastewater.[21] Disinfection can render bacteria non-culturable but does not necessarily render them non-viable.

Such work is of value in determining the environmental significance of bacterial pathogens, particularly where the environment may act as a reservoir for disease. More work needs to be done to understand the viable but non-culturable state and its significance to our present water monitoring techniques based upon cultural methods.

The bacteria that grow in drinking water A wide variety of bacteria are able to grow in drinking water. The main genera include *Pseudomonas* spp., *Aeromonas* spp., *Bacillus* spp., *Acinetobacter* spp., *Alkaligenes* spp., coliforms and micrococci. Where growth occurs it may cause a visible turbidity or a visible flocculated suspension. More important is the growth associated with water distribution systems and the accumulation of a biofilm on internal surfaces. Temperature, available organic material from fittings and the supply water and the degree of stagnation are all important factors. Biofilms will accumulate in the presence of 0.8mg/l of free chlorine and considerable accumulations may occur at concentrations of less than 0.2mg/l.[22]

Biofilms protect microorganisms from the effects of disinfectants and may detach and produce turbid waters. They can cause taste and odour problems, excessive corrosion, and can provide nutrients for the growth of other microorganisms such as amoebae and *Legionella* spp.. The efficiency of disinfectants on opportunist pathogens such as *Pseudomonas aeruginosa* in biofilms can be severely restricted. *Pseudomonas aeruginosa* was found to be able to recolonise PVC pipes after exposure of biofilms to a variety of disinfectants including 10-50mg/l free chlorine for seven days.[23] The growth of coliforms will of course lead to erroneous assumptions about drinking water quality.

One organism which has been isolated from drinking water systems is *Aeromonas hydrophila*. A study in Belgium detected aeromonads in a wide variety of waters, both raw and treated, and resulted in the instigation of routine monitoring programmes.[24] Problems in the United Kingdom have been identified[25] and the public health significance of *Aeromonas* spp. questioned. *Aeromonas* spp. were isolated from non-chlorinated and chlorinated drinking water supplies in Australia.[26,27] With the chlorinated supply, the incidence of *Aeromonas gastroenteritis* paralleled the pattern of isolation of *Aeromonas* spp. in water in the distribution system. *Aeromonas sobria* was also isolated from chlorinated drinking water in the United States.[28] Aeromonads and heterotrophic plate count (HPC) have been shown to be inhibited by copper ions present in some sampled waters, and the incorporation of ethylene-diamine tetraacetic acid (EDTA) into routine sample bottles has been recommended.[29]

If rainfall is decreasing and water temperatures and organic nutrients are increasing, aeromonads could become a problem of the future not just because of their more frequent isolation by membrane filtration, but also because of the increasing concern about their public health significance. Some aeromonads isolated in 1990 in the authors' laboratory produced small yellow colonies on membranes at 44°C whilst failing to ferment lactose at 37°C.

The viruses

Human pathogenic viruses originating from the gastrointestinal tract of animals and man are widely disseminated throughout the environment. Viral agents including Rotavirus, Norwalk Agent, Small Round Viruses, Hepatitis A, Hepatitis Non A - Non B and the enteric adenoviruses 40 and 41 have all been implicated in waterborne outbreaks of disease. Recorded cases of viral gastroenteritis have been known to spread rapidly through a population, involving several thousand cases in a single outbreak. A waterborne outbreak believed to be caused by Non A - Non B Hepatitis was recorded in Delhi, India in 1955-56 involving thirty thousand cases of hepatitis.[30] The health risks associated with viruses in the environment extends beyond recorded cases of gastroenteritis. Coxsackie B virus (CVB) can cause a wide spectrum of diseases. Coxsackie B4 has been implicated in insulin dependent diabetes melitis and is thought to be responsible for the organ specific autoimmune disorder which destroys insulin producing β-cells in the pancreas.[31] Other syndromes associated with CVB are myealgic encephalomyelitis (ME), acute myopericarditis and dilated cardiomyopathy, a chronic cardiac disease which is the second most common reason for cardiac transplants in the United Kingdom.[32] Viruses can therefore cause considerable long term damage to health.

Survival of viruses in the environment Virus survival depends upon a variety of biotic and abiotic factors. Temperature, suspended solids concentration, pH, salinity, ultra-violet light penetration, organic compounds, virus type, ability to aggregate and the presence of bacteria, algae and protozoa can all influence infectious capabilities.[33] Enteroviruses are rapidly inactivated at temperatures exceeding 50°C. Suspended solids provide a certain amount of protection. Absorption on to organic matter with subsequent sedimentation prevents inactivation by ultra-violet light and viruses may survive for long periods in sediments and be resuspended in turbulent conditions. Viruses in surface waters can result in outbreaks from inadequately treated water.[34] Contamination of groundwater by septic tanks, landfill and disposal of sludge to land may represent another serious source of water contamination.

Virus removal in soil relies mainly on adsorption and is not as significantly affected by natural filtration or sedimentation as bacteria. Viruses are readily adsorbed by clay soils and to a lesser extent by loams, but solid associated viruses remain infective and protected from environmental stresses. They present a significant problem in the persistence and transmission of viruses. Sand and gravel do not achieve good removal of viruses whilst limestone fissures can allow viral transport over great distances. In studies, adsorption of different strains were observed on various soil types. It was found that removal of viruses in sandy loam soil varied from 0 to 99.9 per cent depending on virus strain. Overall Echo 1, 12 and 29, Simian rotavirus and Coxsackie B4 remained more dissociated from solids. Groundwater supplies are a valuable source of drinking water often requiring only chlorination minimal treatment to achieve water quality standards. However, the low infective dose of viruses and their high resistance to chlorine leads to concern over their presence in groundwater.

Virus assay Tissue culture assay techniques are used to isolate viruses and although this is expensive, the techniques are relatively simple and results can be obtained within 24-48 hours. Advancements in immunochemistry, tissue culture and genetic engineering have seen the development of more rapid techniques for the identification of viruses. Direct enzyme linked immunosorbent assay (ELISA) and radioimmune-assay (RIA) will produce results within 1-2 hours but their limit of detection is 1,000 plaque forming units (pfu). Immunofluorescence, although more sensitive, is incapable of detecting unculturable viruses and requires a 24-48 hour incubation period.[35]

Recent research into the use of gene probes for environmental monitoring has introduced the possibility of a fast and sensitive assay which may assist or even replace current methodology. Gene probes (pieces of nucleic acid which hybridise with the complementary base pairs of specific genes) can be used to identify the genetic information of any organism. The sensitivity of this method is comparable to that of tissue culture. Gene probes already exist for Rotavirus, the Enteroviruses (including Hepatitis A) and Adenoviruses with the possibility of using a Calcivirus probe to detect Norwalk Agent. These probes are labelled with radioisotopes but it should be possible to combine them with biotin and then use immunofluorescence or ELISA for detection.[35] Gene probes do not rely on tissue culture and are therefore a cheaper alternative. They can be engineered to detect a large group of viruses and could therefore give a general screening of water quality.

Gene probes cannot differentiate between infective and non-infective particles and, since tissue culture will only detect a small proportion of Enteroviruses, counts obtained with gene probes will always be higher. Hepatitis A, which requires 3-6 weeks incubation in tissue culture, is extremely resistant in the environment and gene probes would provide an ideal detection method.

Microbiological indicators for viruses The use of bacterial indicators such as faecal and other coliforms and faecal streptococci has not been directly correlated with incidents of waterborne disease.[7] Bacteriophages have also been proposed as indicators of virus contamination. The advantages are similar to those of the bacterial indicator model. It is however difficult to select one bacteriophage which will be representative of all human viruses. Comparative studies have shown that Enteroviruses may be detected in clean waters in the absence of the bacteriophages O X 174, f2, MS2 and T2.[36] The bacteria and the bacteriophage model cannot be extrapolated to cover all possible microorganisms in all possible situations.

Virological analysis can play an important role in monitoring raw water, treatment processes and water in distribution systems to provide valuable data on the efficiency of existing and prospective operations and provide the necessary information for a more accurate assessment of risk to public health.

The intestinal parasites

Both *Giardia* and *Cryptosporidium* have caused waterborne outbreaks in the United States. A number of waterborne outbreaks of *Cryptosporidium* have occurred in the United Kingdom[32] but only one recorded outbreak of *Giardia*.[37] Epidemiological evidence suggested water, although there was no positive proof of the source nor any confirmed identification of *Giardia* cysts in the water. Methods have now been developed for the detection of *Cryptosporidium* and *Giardia* in waters and water associated materials. Although recovery efficiencies may be low and methodology time consuming, they are being used to acquire knowledge of the incidence and distribution of these organisms in the environment and the effect of water treatment on their removal. Ultimately the knowledge acquired should prevent further outbreaks.

Other protozoa Protozoa have an important role to play in water microbiology. They occur widely in the natural environment. They can and do ingest bacteria. The protozoa may provide a means for bacterial replication and a means for protection from inactivation by chlorine. Protozoa have been investigated as a possible source of the spontaneous appearance of coliforms in chlorinated drinking water.[38] Coliforms were killed by 0.25mg/l of chlorine, but once ingested by protozoa they were observed to survive four to ten times that concentration. *Legionella pneumophila* can survive and multiply within amoebic trophozoites.[39] Infected trophozoites can protect the bacterium from at least 50mg/l of chlorine. *Legionella pneumophila* serogroup 1 was found in 12 per cent of drinking water sources in England.[40] The majority of isolates were found to be viable but non-culturable, could survive conventional water treatment and disinfection but could still retain the ability to colonise warm water systems in buildings. *Legionella* have been isolated from surface waters by culture of amoebae and demonstration of the bacterium in amoebal lysates by immunofluorescence.[41] The investigation was prompted by several cases of legionellosis after near-drowning in the river Rhine.[42] A wide variety of free living amoebae were isolated from surface waters. Some free-living amoebae are pathogenic. *Naegleria* spp. and *Acanthamoeba* spp. can cause primary amoebic meningoencephalitis and keratitis respectively. The latter gives concern because of association with people who wear contact lenses. Such concern has led to the publication of methods for their detection.[43]

Surface water quality

Surface water provides a valuable resource for drinking water as well as recreation. Sudden deterioration following pollution can render an otherwise adequate treatment inadequate. Reported farm pollution incidents have increased steadily from 1,484 in 1979 to 4,141 in 1988, the highest ever recorded.[44] For 1988, fifty-five percent of the incidents were related to cattle slurries. Problems arise from the containment of as much as half a million gallons of slurry during a wet winter. Consideration in the report is given to the chemical contamination of surface water but no mention is made of the microbiological implications. The spreading of slurry and manure has also resulted in infections in humans.[9,45]

Discussion

Much effort is given to the collection, treatment and protection of water supplied for drinking. That few problems occur compared to the amount of water that is supplied reflects both this effort and the quality of water that is produced. There is however no room for complacency. Any reduction in vigilance will result in deterioration of water quality and the possibility of waterborne disease. Part of the vigilance involves being aware of the current and future microbiological issues concerning water quality.

Changes in climate and rainfall will result in changes in the microbiological quality of water, both raw and in distribution. The growth of organisms such as coliforms and aeromonads may be seen more frequently. Newly recognised organisms may be identified. Methods are continually changing and improving and with them our knowledge of the incidence and distribution of microorganisms. Alternative treatments for water may solve one particular problem but give rise to another. Novel forms of water use may allow the growth of particular pathogens. All these issues need consideration if public health is to be protected. The microbiologist needs to keep in step with, even ahead of the microorganisms being chased.

The opinions expressed in this paper are those of the authors and not necessarily those of Yorkshire Water or any other Water Company.

References

1. *Operational Guidelines for the Protection of Drinking Water Supplies,* (1988). Water Authorities Association and Water Companies Association, London.
2. Dadswell, V.J., (1990). 'Microbiological Aspects of Water Quality and Health'. Institution of Water and Environmental Management, Annual Symposium, 5.1 - 5.11.
3. Galbraith, N.S., Barrett, N.J. and Stanwell-Smith, R., (1987). 'Water and Disease after Croydon: A Review of Water-borne and Water-associated disease in the UK 1937 - 1986'. *Journal of the Institution of Water and Environmental Management,* 1, (7), pp 7-21.
4. Benton, C., Forbes, G.I., Paterson, G.M., Sharp, J.C.M. and Wilson, T.S., (1989). 'The Incidence of Waterborne and Water-associated Disease in Scotland from 1945 - 1987'. *Water Science and Technology,* 21, (3), pp 125-129.
5. Hayes, E.B., et. al., (1989). 'Large Community Outbreak of Cryptosporidiosis Due to Contamination of a Filtered Public Water Supply'. *The New England Journal of Medicine,* 320, pp 1372-1376.
6. Tao, H., et. al., (1984). 'Waterborne Outbreak of Rotavirus Diarrhoea in Adults in China Caused by a Novel Rotavirus'. *The Lancet,* i, pp 1139 - 1142.
7. Hejkal, T.W., Keswick, B., Labelle, R.L., Gerba, C.P., Sanchez, Y., Dreesman, G., Hafkin, B. and Melnick, J.L., (1982). 'Viruses in a Community Water Supply Associated with an Outbreak of Gastroenteritis and Infectious Hepatitis'. *Journal of the American Water Works Association,* 74, pp 318-321.
8. Craun, F., (1986). *Waterborne Disease in the United States.* CRC Press, Boca Raton, USA.
9. Smith, H.V., Patterson, W.J., Hardie, R., Greene, L.A., Benton, C., Tulloch, W., Gilmour, R.A., Girdwood, R.W.A., Sharp, J.C.M. and Forbes, G.I., (1989). 'An Outbreak of Waterborne Cryptosporidiosis Caused by Post-treatment Contamination'. *Epidemiology and Infection,* 103, pp 703 -715.
10. 'The Bacteriological Examination of Drinking Water Supplies' (1982). *Reports on Public Health and Medical Subjects No. 71.* Department of Health and Social Security, HMSO, 1983.

11. *Guidance on Safeguarding the Quality of Public Water Supplies*, (1989). Department of the Environment, Welsh Office, HMSO, London.
12. Edberg, S.C., Allen, M.J., Smith, D.B. and the National Collaborative Study, (1988). 'National Field Evaluation of a Defined Substrate Method for the Simultaneous Enumeration of Total Coliforms and *Escherichia coli* from Drinking Water: Comparison with the Standard Multiple Tube Fermentation Method'. *Applied and Environmental Microbiology,* 54, pp 1595 - 1601.
13. Lewis, C.M. and Mak, J.L., (1989) 'Comparison of Membrane Analysis Filtration and Autoanalysis Colilert Presence - Absence Techniques for Analysis of Total Coliforms and *Escherichia coli* in Drinking Water Samples'. *Applied and Environmental Microbiology*, 55, (12), pp 3091 - 3094.
14. Adams, M.R., Grub, S.M., Hamer, A. and Clifford, M.N., (1990). 'Colorimetric Enumeration of *Escherichia coli.* Based on β-glucuronidase activity'. *Applied and Environmental Microbiology*, 56, (7) pp 2021 - 2024.
15. Barnes, R., et. al., (1989). 'Evaluation of the 7 - h Membrane Filter Test for Quantitation of Faecal Coliforms in Water'. *Applied and Environmental Microbiology,* 55, (6), pp 1504 - 1506.
16. Berg, J.D. and Fiksdal, L., (1988). 'Rapid Detection of Total and Faecal Coliforms in Water by Enzymatic Hydrolysis of 4-Methylumbelliferone - β - D - Galactoside'. *Applied and Environmental Microbiology*, 54, (8), pp 2118 -2122.
17. Jones, F. and Watkins, J., (1985). 'The Water Cycle as a Source of Pathogens'. *Journal of Applied Bacteriology Symposium Supplement*. pp 27S-36S.
18. Grimes, D.J. and Colwell, R.R., (1986). 'Viability and Virulence of *Escherichia coli* Suspended by Membrane Chamber in a Semitropical Ocean Water'. *FEMS Microbiology Letters*, 34, pp 161-165.
19. Brayton, P.R., Tamplin, M.L., Huq, A. and Colwell, R.R., (1987). 'Enumeration of *Vibrio cholerae* 01 in Bangladesh Waters by Fluorescent Antibody Direct Viable Count'. *Applied and Environmental Microbiology*, 53, (12), pp 2862-2865.

20. Huq, A., Colwell, R.R., Rahman, R., Ali, A., Chowdury, M.A.R., Parveen, S., Sack, D.A. and Russek-Cohen, E., (1990). 'Detection of *Vibrio cholerae* 01 in the Aquatic Environment by Fluorescent-Monoclonal Antibody and Cultural Methods'. *Applied and Environmental Microbiology*, 56, (8) pp 2370-2373.

21. Desmonts, C., Minet, J., Colwell, R. and Corimer, M., (1990). 'Fluorescent-Antibody Method Useful for Detecting Viable but Nonculturable *Salmonella* spp. in Chlorinated Wastewater'. *Applied and Environmental Microbiology*, 56, (5), pp 1448 - 1452.

22. Van der Wende, E., Characklis, W.G. and Smith, D.B., (1989). 'Biofilms and Bacterial Drinking Water Quality'. *Water Research,* 23, (10), pp 1313 - 1322.

23. Anderson, R.L., Holland, B.W., Carr, J.K., Bond, W.W. and Favero, M.S., (1990). 'Effect of Disinfectants on Pseudomonads Colonised on the Interior Surface of PVC Pipes'. *American Journal of Public Health,* 80, (1), pp 17-21.

24. Meheus, J. and Peters, P., (1989). 'Preventative and Corrective Actions to cope with Aeromonas Growth in Water Treatment'. *Water Supply*, 7, P10-1-P10-4.

25. Edge, J.C. and Finch, P.E., (1987). 'Observations on Bacterial Aftergrowth in Water Supply Distribution Systems: Implications for Disinfection Strategies'. *Journal of the Institution of Water and Environmental Management*, 1, (7), pp104-110.

26. Burke, V., Robinson, J., Gracey, M., Peterson, D., Meyer, N. and Haley, V., (1984). 'Isolation of *Aeromonas* spp. from an Unchlorinated Drinking Water Supply'. *Applied and Environmental Microbiology,* 48, (2), pp 367 - 370.

27. Burke, V., Robinson, J., Gracey, M., Peterson, D. and Partridge, K., (1984). 'Isolation of *Aeromonas hydrophila* from a Metropolitan Water Supply : Seasonal Correlation with Clinical Isolates'. *Applied and Environmental Microbiology*, 48, (2), pp 361-366.

28. Lechevallier, M.W., Evans, T.M., Seidler, R.J., Daily, O.P., Merrell, B.R., Rollins, D.M. and Joseph S.W., (1982). *'Aeromonas sobria* in Chlorinated Drinking Water Supplies'. *Microbial Ecology*, 8, pp 325 -333.

29. Versteegh, J.M.F., Havelaar, A.H., Hoekstra and Visser, A., (1989). 'Complexing of Copper in Drinking Water Samples to Enhance Recovery of *Aeromonas* and Other Bacteria'. *Journal of Applied Bacteriology*, 67, (5), pp 561-566.
30. Khuroo, M.S., (1980). 'Study of an Epidemic of non-A, non B Hepatitis. Possibility of another Human Hepatitis Virus Distinct from Post-transfusion non A, non B Type'. *American Journal of Medicine*, 68, pp 818-824.
31. Toniolo, A., Federico, G., Basolo, F., Onodera, T., (1988) 'Diabetis Mellatis' in *Coxsackie Viruses - A general update*. Friedman H (ed) Plenum Press, London.
32. Muir, P., Tilzey, A.J., English, T.A.H., Nicholson, F., Signey, M. and Baratvala, J.E., (1989). 'Chronic Relapsing Pericarditis and Dialated Cardiomyopathy. Serological Evidence of Persistent Enterovirus Infection'. *The Lancet*, i, pp 804-806.
33. Geldenhuys, D.C., and Pretorius, P.D., (1989). 'The Occurrence of Enteroviruses in Polluted Water, Correlation to Indicator Organisms and Factors Influencing their Numbers'. *Water Science and Technology*, 21, (3), pp 105-109.
34. Rose, J.B., (1990). 'Emerging Issues for the Microbiology of Drinking Water'. *Water / Engineering and Management*, July, pp 23-29.
35. Gerba, C.P., Margolin, A.B. and Hewlett, M.J., (1989). 'Application of Gene Probes to Virus Detection in Water'. *Water Science and Technology*, 21, (3), pp 147-154.
36. Keswick, B.M., (1982). 'Monitoring Disinfection of Enteric Viruses with Bacteriophage'. *Viruses and Disinfection of Water and Wastewater* pp 298 - 305, University of Surrey.
37. Browning, J.R. and Ives, D.G., (1987). 'Environmental Health and the Water Distribution System : A Case History of an Outbreak of Giardiasis'. *Journal of the Institution of Water and Environmental Management*, 1, (7), pp 55-60.
38. Shotts, E.B. and Wooley, R.E., (1990). *Protozoan Sources of Spontaneous Coliform Occurrence in Chlorinated Drinking Water. Project Summary*. United States Environmental Protection Agency, Project No. EPA/600/52-89/019. Cincinnati, OH 45268.
39. Kilvington, S. and Price, J., (1990). 'Survival of *Legionella pneumophila* Within Cysts of *Acanthamoeba polyphaga* following Chlorine Exposure'. *Journal of Applied Bacteriology*, 68, pp 519-525.

40. Colbourne, J.S. and Dennis, P.J., (1987). 'The Ecology and Survival of *Legionella pneumophila.*' *Journal of the Institution of Water and Environmental Management*, 1, (7), pp 345-350.
41. Harf, C. and Monteil, H., (1989). 'Pathogenic Microorganisms in Environmental Waters : A Potential Risk for Human Health'. *Water International*, 14, (2), pp 75-79.
42. Hasselmann, M., Lutun, P., Schneider, F., Harf, C., Kieffer, P., Monteil, H. and Tempe, J.D., (1987). 'Maladie des Legionnaires apres Noyade dans le Rhin'. Medecine et Maladies Infectieuses, *France,* 12, p752.
43. *Isolation and Identification of Giardia Cysts, Cryptosporidum Oocysts and Free Living Pathogenic Amoebae in Water etc.,* (1989). Standing Committee of Analysts, London, HMSO.
44. *Water Pollution from Farm Wastes*, (1988). England and Wales. A Joint Report from the Water Authorities Association and the Ministry of Agriculture, Fisheries and Food.
45. Casemore, D., (1989). 'The Epidemiology of Human Cryptosporidiosis'. *PHLS Microbiology Digest,* 6, (2), pp 544-66.

4. Advanced treatments for better drinking water quality

J. Mallevialle and J.P. Duguet

Introduction

This chapter addresses possible solutions to problems (eg. microbiological parameters, nitrates, pesticides, tastes and odours) arising from the application of European standards. It also deals with the side effects of each solution, such as the formation of oxidation and disinfection by-products, reagent residuals, etc. A technico-economic survey of the different solutions is then presented. The survey focuses firstly on the optimisation of the present treatment lines (clarification, oxidation, adsorption on activated carbon, disinfection), and secondly, on the use of recent techniques such as oxidant coupling (ozone/hydrogen peroxide).

Throughout the world, the water distribution profession has experienced rapid change, partly because of the strategy adopted by the water industry, and partly as a result of external influences such as increasingly strict quality standards, and growth in various industrial sectors (electronics, biotechnology, etc), the reduction in the quantity and quality of water resources, and economic pressures.

The principal quality-related problems that France faces as a result of the directive covering application of the new European standards, are microbiological parameters, pesticides, tastes, odours and nitrates.[1,2] This chapter seeks to demonstrate that it is possible to overcome these problems today provided that additional investments are made. This short chapter does not claim to offer an exhaustive review of these possibilities, which could have involved developing different scenarios that take into account the size of the plants and the location of those plants in urban or rural areas.

Microbiological parameters

In view of the fact that the failure to meet standards is not generally a permanent state of affairs, the basic problem here is one of reliability. This clearly involves increasing the reliability of the treatment processes employed. The research that has been conducted over the last two years into disinfection by ozone and by chlorine derivatives (kinetics of reaction, effects of competition, hydraulic design of reactors) has demonstrated the need to review current recommendations. It is no longer sufficient to maintain a stated level of residual disinfectant for a specified time. The notion of CT must be accepted, where C is the concentration of disinfectant in the water and T is the real hydraulic retention time rather than the average retention time (which does not account for the fact that flows within the tank are not uniform). Applying the notion of CT should lead to significant changes in the design of reactors or contact tanks. To achieve high reliability, disinfection processes must also remove all suspended particles, partly because filtration is still the most effective way of removing certain pathogenic microorganisms such as *Giardia Lamblia*, and partly because a large number of germs become fixed to these suspended particles and the action of disinfectants is thus nullified. This aspect of the problem has been overcome in the drinking water plants in the Paris area that are equipped with sophisticated treatment lines (clarification, ozonisation, granular activated carbon filtration, post-disinfection), but is considerably more difficult to solve at the other end of the spectrum, such as, in small-scale plants located in rural areas, where the most reliable solution would undoubtedly be filtration on micro- or ultrafiltration membranes.[3]

However, it is not sufficient to guarantee perfect microbiological quality of the water that leaves the drinking water plants. It must still be conveyed to the consumer, and that requires a well designed and well maintained distribution network, with an interconnected architecture to avoid excessive retention times and judicious choice of materials to minimise corrosion and biofilm formation. But even these conditions are insufficient if upstream treatment is incomplete. Problems can arise from scale formation or agressivity, suspended particles due to ineffective clarification techniques, post-precipitation of iron or manganese, and presence of carbon that can be assimilated by micro-organisms. The problem can be solved partly by maintaining a level of residual disinfectant in the distribution network, but this approach will not be really reliable unless operators have a close awareness of the hydraulic properties of the network and the kinetics of consumption of the disinfectant, and unless concentrations of residuals are measured on a virtually permanent basis. Recent developments in computerised mapping are liable to increase significantly the reliability of this kind of solution, at the same time as minimising the formation of disinfectant by-products such as trihalomethanes (Societe Parisienne des Eaux is currently conducting a study in this area).[4]

Nitrates

Recent publications have dealt so extensively with the nitrates problem that it would appear superfluous here to dwell upon the fact that the fight against nitrates should begin at the source of the pollution. As far as drinking water production plants are concerned, two treatment processes have been approved by the French Ministry of Health. These are denitration using ion-exchanging resins, which is the most widespread physico-chemical process employed at this time but generates eluates that contain high concentrations of nitrates and sodium chloride, and biological denitrification which converts the nitrate ion into nitrogen gas. Several denitrification techniques have been developed, based either on the use of autotrophic bacteria in the presence of sulphur, or on heterotrophic bacteria which require additional carbon (acetic acid or ethanol). These biological processes must be combined with downstream carbon filtration and disinfection. A few production units using these processes have been operating for five or six years, such as the 50m^3/hr unit at Chateau Landon (Seine et Marne).[5]

Pesticides

The residual pesticides problem is more complex. Maximum concentration level (MCL) values are very low (100 ng/l or less), and the range of substances involved is enormous. In 1985, out of approximately 100,000 tons of pesticides used in France, insecticides accounted for around 6,000 tons (organohalogenated, organophosphates, carbamates); fungicides accounted for around 50,000 tons (sulphur, carbamates, copper) and herbicides for 35,000 tons (triazines, chlorophenoxyacetic acid, and urea substitutes).[1] Among the substances that can be analysed in terms of MCL, only atrazine and simazine (triazine family) pose a problem, and this is valid throughout Europe. Concentrations measured in France are between 10 nanograms and a few micrograms per litre. Few treatment processes are capable of removing these compounds with the 90 per cent to 99 per cent efficiency required in certain instances. Adsorption on activated carbon is an effective solution, but it should be remembered that the effects of competition with the organic matrix of the water, the fact that concentrations of this type of pollutant vary considerably with time, and the lack of sensors for continuous monitoring or fast analysis techniques lead to oversizing of facilities as a safety measure. For example, in surface water such as Seine water, which would contain between 1 and 2 micrograms of pesticides per litre, activated carbon filters would need to be regenerated every three months, although these filters can effectively remove tastes and odours for up to several years.

Recent research has shown that hydrogen peroxide/ozone coupling is particularly effective, whilst the use of ozone alone is only marginally useful. This process generates highly reactive oxygenated radicals such as OH hydroxyl radicals and is relatively easy to introduce into plants that already have ozonisation equipment. Peroxide addition can be started and stopped quickly to handle periods of high pollution.[6] Industrial-scale tests are underway at the Lyonnaise des Eaux facility at Le Pecq and at Le Mont Valerien (Compagnie des Eaux de Banlieue -CEB).

Various techniques have been developed recently or are still under development. One of these techniques involves coupling powdered activated carbon with ultrafiltration membranes in an integrated unit. This approach has demonstrated its effectiveness for removing suspended particles, microorganisms and trace pollutants such as pesticides, chlorinated solvents and the molecules that cause tastes and odours.[7] A 10 m^3/hr demonstration unit is operating at Le Mont Valerien (CEB).

Tastes and odours

Removing tastes and odours still relies on a mixture of experience and more scientific approaches. It is the most complex quality related problem to understand and to solve. This complexity is the result of several factors. Numerous types of taste and odour descriptors (hydrocarbons, medicinal substances, etc) can appear, for example, and traces (nanograms or a few micrograms per litre) of numerous organic compounds can interact in different ways and cause very temporary problems.[8]

Tastes and odours can be generated at the source, during the production process, or in the distribution network. Solutions to tastes and odours caused at source are easy to list, but are often very difficult to apply because the water distributor is generally not responsible for anomalies at source level. Source problems include eutrophication and accidental pollution. As regards problems originating in the production plant, it is important to be aware that each potential process can have negative as well as positive effects. The negative effects are caused by the oxidation by-products that are formed during the reaction between ozone or chlorine derivatives and the organic matrix of the water, or simply by the presence of residual reagents such as chlorine. The many solutions that exist for this type of problem are too numerous to mention in detail here, but there is no doubt that the most effective way of minimising the probability of tastes and odours appearing in drinking water during the production process involves coupling an oxidant (ozone alone or combined with oxygenated water) with an adsorbant (granular activated carbon). For small-capacity plants, coupling powdered activated carbon with ultrafiltration membranes has a number of significant advantages. Solutions to problems of tastes and odours originating in distribution networks are similar to those recommended for microbiological parameters.

Economic aspects

Precise capital expenditures and operating costs can only be estimated on a site-by-site basis. These expenditures are heavily dependent on the type of source (surface water or ground water), the type of existing treatment processes, the size of the plant, and the interaction between the different problems that need to be solved. By way of example, minimum and maximum costs are indicated in Tables 1 to 3 below for a 3,000 m^3/hr plant (unless otherwise stated) for a range of treatment objectives.

Table 1

OBJECTIVE : low concentrations of pesticides, tastes and odours SOLUTION : powdered activated carbon			
Costs (FF/m^3)	Capital Expenditure	Operation	Total
dose: 30 to 50 g/m^3	0.02	0.18 to 0.30	0.20 to 0.32

Table 2

OBJECTIVE : low concentrations of pesticides, tastes and odours, carbon assimilable by microorganisms, microbiological parameters SOLUTION : ozone/granular activated carbon coupling			
Costs (FF/m^3)	Capital Expenditure	Operation	Total
assimilable carbon tastes and odours, microorganisms	0.55 to 1.10	0.15	0.70 to 1.25
additional cost for pesticides	0.01	0.13	0.14

Table 3

OBJECTIVE : low concentration of nitrates SOLUTION : ion exchange or biological reactor			
Costs (FF/m^3)	Capital Expenditure	Operation	Total
ion exchange 10 to 50m^3/hr	1.20 to 3.10	0.9 to 1.20	2.10 to 4.30
biological reactor	0.78	0.9	1.68

Comments on Tables 1 to 3

* Capital expenditures expressed in FF/m^3 are based on amortizement over 20 years.

* The powdered activated carbon solution is only effective when tastes and odours are present at low intensities.

* Coupling ozone with activated carbon filtration offers other benefits: removal of trihalomethanes formation potential, buffer capacity against accidental pollution, etc.
* 'Additional cost for removing pesticides' involves increasing the frequency of regeneration of the granular activated carbon, injecting larger quantities of ozone, and adding hydrogen peroxide.
* The cost of denitration is calculated on the basis of raw water containing 85 mg/l and treated water containing 15 mg/l. For biological denitrification, these values are 70 and 15 mg/l respectively.
* Due to lack of long term data about the operation of industrial-scale plants, no reference to membrane processes is made in the tables.

Conclusion

With the exception of nitrates, all the problems described in this paper are complex. The complexity is a result of the fact that pollutants or microorganisms exist in very low concentrations and must be separated or removed from an organic and mineral matrix without causing excessive negative side effects. Because of the competition involved in adsorption processes and in oxidation or disinfection processes, and because the concentrations of these pollutants vary greatly with time and cannot be measured on a continuous basis because sensors offering the requisite sensitivity and reliability do not exist, there is a need to use overdosages of reagents and thus to oversize installations in order to achieve maximum reliability. The introduction of more stringent drinking water standards must be accompanied by significant research efforts focusing on characterisation of the organic matrix of the water and the related formation of by-products, the development of sensors or fast analysis techniques, and the development of innovative treatment processes such as the use of membranes. These have the particular advantage that they are absolute filters and obviate the need to add reagents. This process is already underway, and will probably lead to a rise in the price of water in the foreseeable future. Nevertheless, these efforts will be fruitless if water source quality does not improve at the same time, and that relies on removing pollution before it is released into the environment.

References

1. Decree no. 89-3 of January 3, (1989).
2. Mallevialle, J. and Chambolle, T., 'La Qualite de l'Eau', *la Recherche* no. 221, (1990), pp. 598-606.
3. Bersillon, J.l., Anselme, C., Mallevialle, J., Aptel, P. and Fiessinger, F., (1989) 'Ultrafiltration Applied to Drinking Water Treatment: the Case of Small Systems', 7th ASPAC-IWSA Water Nagoya.
4. Wable, O. et. al., (1990) 'Modelling Chlorine Concentration in Distribution Networks', Water Quality Technology Conference (AWWA), November 11-15, San Diego, USA.
5. Degremont, (1989)*Water Treatment Handbook*, volumes I and II, Lavoisier.
6. Duguet, J.P., Anselme, C. and Mallevialle, J. (1989) 'New Advances in Oxidation Processes: Some Examples of Application of Ozone/Hydrogen Peroxide Combination for the Removal of Micropollutants from Drinking Water', 7th ASPAC-IWSA Water Nagoya.
7. Anselme, C. and Charles, P. (1990) *The use of Powdered Activated Carbon for the Removal of Specific Pollutants in Ultrafiltration Processes*, Prix Chemviron.
8. Mallevialle, J. and Suffet, I.H. (1987) *Identification and Treatment of Tastes and Odours in Drinking Water*, American Water Works Association, Denver Co., USA.

5. Lessons learnt from waterborne outbreaks of cryptosporidiosis

C. Benton

In April 1988 the first documented UK waterborne outbreak of cryptosporidiosis occurred in Ayrshire, Scotland, in which oocysts were detected in treated water. Extensive epidemiological investigations and the close collaboration which took place between the many disciplines involved in the investigation helped to establish the waterborne nature of this outbreak.

Prior to 1988 two documented outbreaks of waterborne cryptosporidiosis had occurred in the USA. Valuable information was obtained from the experiences gained from these outbreaks which helped Strathclyde Water in their investigations.

In February 1989 the second UK waterborne outbreak of cryptosporidiosis occurred in the Swindon/Oxfordshire area supplied by Thames Water. This outbreak prompted the government to set up an expert group, chaired by Sir John Badenoch to advise them on the significance of cryptosporidium in water supplies.

Many valuable lessons have been learnt for the water industry and others, both from the experiences gained from waterborne outbreaks and from the deliberations of the Badenoch group of experts. These lessons are outlined and detailed in this chapter.

Introduction

In March 1989, following an outbreak of cryptosporidiosis in Swindon and Oxfordshire, the Secretary of State for the Environment announced that an expert group was to be set up in consultation with the Department of Health, under the chairmanship of Sir John Badenoch, one of Britain's most distinguished gastro-intestinal specialists. The expert group was given the remit of advising the Government on the significance of Cryptosporidium in water supplies. An interim report was produced by the expert group in July 1989 and the final report in July 1990.[1]

Cryptosporidium (Greek hidden spore) is primarily an enteric protozoan parasite which was for many years considered to be predominantly a pathogen of livestock and other mammals. It is only in the last decade that is has become evident that Cryptosporidium parvum can be a significant pathogen of human beings. Cryptosporidiosis can be spread directly to man from animals (cattle and sheep, with calves and lambs being an important reservoir of infection). Person to person spread is also an important means of transmission particularly within families, playgroups, nursery schools, day care centres, hospitals and other institutions. Cryptosporidium is also emerging as an important cause of travellers' diarrhoea. Specific foods such as raw sausages, tripe, raw milk have been associated with cryptosporidiosis[2,3] but there is no evidence to suggest that such contamination is a significant route of infection.

Cryptosporidiosis in the normal healthy individual is characterised by an acute self-limiting diarrhoeal illness, commonly of two to three weeks duration. The disease begins with a sudden onset of gastrointestinal and sometimes 'flu like' symptoms. Cramping abdominal pain, vomiting and loss of appetite are also common. In patients with depressed immunity, including those with AIDS, the disease can be much more serious. Currently there is no specific treatment for cryptosporidiosis, other than supportive treatment only to prevent significant dehydration of the patients.

Evidence for waterborne transmission

Water is also emerging as an important vehicle for the transmission of Cryptosporidium spp. oocysts, as several waterborne outbreaks have now been documented both in the USA (Table 1) and the UK (Table 2). The numbers of reported cases in the UK have risen from a total of 2,139 in 1985, to a total of 9,147 cases reported during 1989.

In 1989 Cryptosporidium was the fourth most commonly identified cause of gastrointestinal infection, accounting for eight per cent of all cases in England and Wales and 13 per cent in Scotland. Although the disease can be waterborne, the proportion of cases infected by this route represents only a very small fraction of all cases of waterborne diarrhoeal illness. Should mains water become contaminated, many cases of illness can occur in the local area of supply.

The characteristics of Cryptosporidium spp. oocysts which aid their spread and survival in the environment are:-

i) they occur as environmentally robust oocysts which are shed in the faeces of an infected host in very large numbers;
ii) the ability of Cryptosporidium parvum oocysts to spread from man to animals and from animals to man;
iii) their resistance to adverse environmental factors;
iv) their ability to remain viable for months in moist soil or up to a year in clean water;
(v) they are unaffected by chlorine in concentrations normally used in water treatment processes and mains disinfection/repair situations.

Documented waterborne outbreaks

USA outbreaks

Table 1 Documented outbreaks of waterborne cryptosporidiosis in the USA

Location	Numbers Affected	Suspected Cause	Oocysts detected in water
Texas 1984	79	Sewage contamination	None
Georgia 1987	13,000 (estimated)	Operational irregularities	2.2/1

Prior to the first UK outbreak, two waterborne outbreaks were documented in the USA, which are summarised in Table 1. The first of these, which occurred in Braun Station Texas in a community of 6,000 people in July 1984, affected 79 individuals.[4] Oocysts were detected in 59 per cent of the stool specimens with an overall attack rate in the community of 34 per cent. Norwalk virus was also detected amongst those infected, which suggested that human sewage may have contaminated the water supply. The cause of this outbreak was thought to be post-treatment contamination of an artesian well supply by sewage. Tracer dye studies supported this hypothesis.

The unfiltered well water received chlorination as the only treatment process. Chlorination alone at the levels used in the treatment of drinking water will not kill C.parvum oocysts. Coliform organisms were routinely detected in the unchlorinated water but never in the chlorinated water. Thus the usefulness of the coliform organism as an indicator of protozoan and viral contamination must be questioned. No cryptosporidium spp. oocysts were detected in the raw or final water although, at the time, the identification techniques for Cryptosporidium spp. oocysts were not well developed.

The second USA outbreak of cryptosporidiosis occurred in Carrollton, Georgia in a community of around 65,000 people in 1987.[5,6] This outbreak affected around 13,000 people (estimated by a retrospective random telephone survey and by serum sampling of exposed residents). Cryptosporidium spp. oocysts were detected in 39 per cent (58/147) of the stool specimens submitted, with an overall attack rate of 40 per cent in the exposed population. No other pathogens were isolated during this outbreak from stool specimens and no indication of concurrent infection was evident from serological studies.

The common water supply to the area of Carrollton was derived from a river fed by a lake and a number of streams. There was a sewage outflow upstream of the catchment area. The water was treated by a conventional two stage process which consisted of coagulation, sedimentation, rapid sand filtration and disinfection with chlorine, with no recycling of backwash water. All treated water samples which were taken from the system were negative for coliform organisms. They also met the USEPA and State of Georgia standards which had been set for turbidity (less than one nephelometric turbidity unit (NTU)) and free chlorine levels (1.5mg/l after disinfection and 0.5mg/l within the distribution system). Although all the statutory standards for this water supply were met, operational difficulties were identified retrospectively, namely:-

(i) The temporary removal of mechanical agitators scheduled for replacement decreased the efficiency of the flocculation step. This resulted in the impaired removal of particulate matter.

(ii) The impairment of the efficiency of filtration, due to the failure of equipment and procedures used to control the flow of water through the filter and the inadequate turbidity monitoring after blending the water. When certain individual filters were monitored, turbidity levels were as high as 3.2 NTU although after blending the average turbidity ranged from 0.4 - 0.8NTU.

(iii) The restarting of filters after shutdown without first backwashing, which could have led to the discharge of dirt, flocculant particles and microorganisms including oocysts from the filter beds into the treated water. Cryptosporidium spp. oocysts were detected in 7 out of 9 samples of treated water and concentrations averaged 0.63 oocysts per litre, with the highest level of 2.2 per litre being found in a 24 hour sample taken post filtration.

The source of this outbreak was never fully determined. There was, however, strong circumstantial evidence through retrospective studies of the population which suggested that there may have been a low level infection with Cryptosporidium spp. oocysts prior to the actual outbreak. The occurrence of Cryptosporidium spp. oocysts in cattle in the watershed was found to be low and the distribution of infected animals did not match the distribution of samples of water which contained Cryptosporidium spp. oocysts. It was possible that the changes in operational practices outlined above allowed a peak of oocysts present in the raw water to pass through the water treatment process in sufficient numbers to cause widespread illness.

In addition, a sewage spill above the treatment plant and an increase in rainfall just after the onset of the outbreak may have increased the load of Cryptosporidium spp. oocysts reaching the plant.

In the USA between 1971 and 1988, 564 outbreaks of waterborne disease have occurred.[7] Cryptosporidiosis has accounted for less than one per cent of documented waterborne outbreaks. The most common cause of outbreaks of waterborne disease was gastroenteritis of undetermined aetiology, followed by giardiasis. When the number of cases of waterborne disease were analysed from the same period 1971 to 1988, cryptosporidiosis accounted for nine per cent of the cases and ranked third in importance after gastroenteritis and giardiasis. Due to the explosive nature of a waterborne outbreak of cryptosporidiosis and the ability of the organism to spread from person to person, the contamination of mains water with Cryptosporidium spp. oocysts can give rise to many cases of illness in the exposed population when compared with other bacterial and chemical agents.

UK outbreaks

The first waterborne outbreak of cryptosporidiosis in the UK in which Cryptosporidium spp. oocysts were detected in treated water occurred in the towns of Saltcoats and Stevenston in Ayrshire, Scotland in April 1988 (Table 2). Many hundreds of people were known to have been affected by the outbreak, although only 27 were diagnosed stool positive, 37 per cent of whom were adults over 20 years. No other pathogens were isolated in the stools of infected persons.

Table 2 Documented outbreak of waterborne cryptosporidiosis in the UK

Location	Numbers Affected	Suspected Cause	Oocysts detected in treated water
Ayrshire 1988	27	Post treatment contamination	4.8/1
Swindon/ Oxfordshire 1989	>500	Source contamination	5/1
Loch Lomond 1989	~244	Under investigation	0.75/1
Nth Humberside	477	Under investigation	None

At the time of the outbreak the drinking water to the towns of Saltcoats and Stevenston were supplied from Camphill treatment works, through storage tanks at the Greenhead works in Stevenston. Camphill treatment works is a modern two stage plant which is treated by a conventional process consisting of coagulation, sedimentation, rapid sand filtration and disinfection with chlorine. The water supply from Camphill to Greenhead works was at the time of the outbreak supplied via transmission pipelines routed through Munnoch Break Pressure Tank.

Camphill supplied around 40,000 people in other parts of Ayrshire where no cases of cryptosporidiosis were reported.

Close collaboration between all the parties investigating this incident was initiated at a very early stage during the outbreak. Extensive sampling, which was intensified after heavy rainfall, helped to establish the cause of the outbreak. The source of the outbreak was found to be an old fire-clay pipe which was discharging run-off from the surrounding catchment area into Munnock Break Pressure Tank. The fireclay pipe was thought to have been sealed off many years before. Cattle slurry and dung had been recently applied to the surrounding land. This outbreak was caused by post-treatment contamination of a trunk main routed through a break-pressure tank into which Cryptosporidium spp. oocysts had been introduced after heavy rainfall.

The second UK outbreak of cryptosporidiosis occurred in Swindon and parts of Oxfordshire in February and March 1989 (Table 2).[10] During this outbreak only 516 cases were diagnosed stool positive, 290 in Swindon and 226 in Oxford.

Thames Water supplied a large part of the affected area in Swindon and Oxfordshire with River Thames water from Farmoor reservoir. This water supplied both the Farmoor and Swinford water treatment works. In each case treatment was by conventional coagulation and rapid sand filtration, followed by disinfection with a final dose of chlorine of 1.5mg/l free chlorine with a minimum of two hours contact time.

Attack rates were generally higher in Swindon than in Oxfordshire. The second highest attack rate occurred in a small community in Oxfordshire which received its water from a third surface waterworks at Worsham. This works abstracted water from the River Windrush, a tributary of the River Thames located upstream of the Farmoor reservoir.

Extensive monitoring was carried out and oocysts were detected in the treated water. Throughout the period of the investigation treated water was free from coliform organisms and no adverse changes were found when turbidity, chlorine demand, coagulant residuals and bacterial plate counts were measured.

Although the cause of the outbreak was never fully established, the findings of the epidemiological and environmental investigations are consistent with an episode of contamination by animal waste, whereby Cryptosporidium spp. oocysts were washed into the River Windrush. The filters at the water treatment works were unable to handle the loading of Cryptosporidium spp. oocysts under these circumstances, despite their providing greater than 99 per cent (2 log) removal capacity.

In Scotland, cryptosporidiosis has been a reportable disease since January 1989. From January to June 1989, 442 cases of cryptosporidiosis were reported. A detailed investigation was carried out in the three health boards in which the highest incidence of 244 cases were reported. The follow up questionnaires upon 206 cases revealed that the onset of symptoms for the majority of cases was April or May and 22 per cent of the cases had been admitted to hospital. Around 39 per cent of the cases had been supplied at home with treated water from Loch Lomond.

Loch Lomond water supplies approximately 20 per cent of the population of Scotland and the water treatment consists of microstraining, pH correction and chlorination. Analysis of this water showed that low levels of Cryptosporidium spp. oocysts were detected in both the raw and final waters.

The discovery of low levels of oocysts in the final water, taken with the epidemiological evidence, points to an association between Loch Lomond water and the incidence of disease. A further study is being undertaken to monitor closely the incidence of cryptosporidiosis in the areas supplied by Loch Lomond water.

In North Humberside, Yorkshire, an outbreak of cryptosporidiosis began in late December 1989. By the end of May when the outbreak was over, there had been 477 laboratory confirmed cases diagnosed stool positive. Mains water from Yorkshire Water Authority's Barmby works was implicated as one of the possible sources of the outbreak. No oocysts were detected in the final treated water. At the time of writing, this outbreak is still being investigated by CDSC, Yorkshire Water and the Drinking Water Inspectorate.

Lessons learnt

The following lists indicate for five important areas of concern, the lessons learnt from the studies of waterborne outbreaks of cryptosporidiosis described above.

Epidemiological investigations

(i) Cryptosporidiosis should be designated as a 'reportable' disease.
(ii) At the earliest stage of a suspected waterborne outbreak, epidemiological investigation should be initiated. It is important to use a standard questionnaire produced by CDSC (England/Wales) or CDS (Scotland) and to map the geographical distribution of cases with the water supply zones.
(iii) During a suspected waterborne outbreak it is desirable to screen all stool specimens for Cryptosporidium spp. oocysts. Evidence from outbreaks in the UK suggest a higher proportion of primary adult cases than would otherwise be expected.

(iv) Livestock are an important reservoir of cryptosporidial infection. Young calves and lambs may be an important source of infection. Veterinary centres should be consulted early in an investigation.
(v) Pet animals do not appear to be an important source of infection. It may be useful to take stool specimens, if only to exclude that route.
(vi) There is no evidence to suggest that food is a significant source of Cryptosporidium spp. oocysts.

Source pollution

(i) It is possible that most of the Cryptosporidium spp. oocysts found in both surface and ground water derive from agricultural practices.
(ii) Agricultural practices have been implicated in at least three of the documented outbreaks.
(iii) Outbreaks have occurred in spring and autumn after periods of heavy rainfall proceeded by dry spells.
(iv) Catchment control practices should be reviewed. There is a need for close consultation to take place between the various undertakers who are responsible.
(v) Cryptosporidium spp. oocysts in environmental waters may also arise from the disposal of products of the sewage treatment process.
(vi) Large numbers of Cryptosporidium spp. oocysts, if present, may pass through the sewage treatment process.
(vii) Sewage sludge, which may be used in agricultural practice or deposited in a designated land tip site, may contain oocysts.

Water treatment practices

(i) There is a risk that even two stage water treatment facilities may not ensure the safety of drinking water supplies. Designers of water treatment works can no longer depend on chlorination to kill all microorganisms.
(ii) Changes in water treatment processes may give rise to less efficient removal of Cryptosporidium spp. oocysts.
(iii) The bypassing of filters and restarting of filters without first backwashing should be avoided.
(iv) Turbidity monitors should ideally be fitted to measure turbidity on each filter.

New Health Considerations in Water Treatment

(v) Coagulant aids should be tested to assess their value in assisting flocculation and retention of oocysts.

(vi) Borehole linings and seals should be maintained to a high standard.

Distribution of water

(i) The normal management of the distribution system should involve no hazard from cryptosporidium spp. oocysts.

(ii) Contamination may occur in an emergency when reduced water pressure allows ground water, which may contain sewage, to penetrate into the distribution system.

(iii) Grazing of grass covered reservoirs with livestock, or the access of livestock to reservoirs, should be prevented at all times.

(iv) Covered service reservoirs are at risk from poor air vents, the means of access and structural cracks.

(v) Routine disinfection with chlorination following an emergency, or during a planned repair and replacement of mains, cannot provide a safeguard against Cryptosporidium spp. oocysts. Operational practices should include provisions for actions which minimise the possibility of contamination.

(vi) Water companies and the Drinking Water Inspectorate should keep under review the existing guidelines for the repair and maintenance of distribution systems. These guidelines should be modified and updated wherever possible.

Liaison

(i) Contingency plans should be made between health authorities, local authorities and water undertakings for dealing with suspected waterborne outbreaks.

(ii) Water undertakings need to review their existing emergency plans for dealing with a suspected waterborne outbreak.

(iii) Individual responsibilities and communication routes need to be set out within the water undertaking and within other organisations.

(iv) An outbreak control team should be established to address the following three areas. Monitoring and sampling, action to deal with the contamination and provision of alternative supplies.

(v) Close communication between the water undertakings, health authorities,

local authorities, CDSC (England and Wales), CDS (Scotland) and all other interested parties should be undertaken.
(vi) Early detection of Cryptosporidium spp. oocysts in water supplies would facilitate the management and control of an outbreak.

Conclusions

1. Further prospective epidemiological studies should be carried out and more funding given to this area.
2. Wherever possible, levels of the pathogenic microorganisms should be reduced at source. The sources of contamination should be identified and their impact upon the catchment area or watershed studied.
3. The removal of microorganisms through current and novel water treatment processes should be evaluated. Treatment works design should be developed to optimise the removal of pathogenic microorganisms through each stage of the process, rather than relying on disinfection as the main barrier.
4. Disinfection guidelines need to be more precisely based and more carefully designed to maximise the efficiency of the unit process.
5. Improved communication is needed between all parties interested in waterborne disease both within and outwith the water industry.
6. The effectiveness of drinking water standards and monitoring and compliance programmes should be evaluated in terms of their ability to prevent waterborne outbreaks. Lessons learnt from outbreaks of waterborne disease should be considered when the regulators are reassessing standards.
7. In the USA, newly identified enteric organisms which have the potential to cause waterborne disease are emerging, examples are Yersinia, E.coli 157, calicivirus, small round viruses and non-A, non-B hepatitis and others. Giardiasis and viral gastroenteritis rank in the USA as the most important causes of waterborne outbreaks, with fifty per cent of outbreaks of unknown aetiology. Epidemiologists, medical microbiologists, water microbiologists, environmental health officers, medical officers and others require to keep abreast of microorganisms which have the potential to cause waterborne disease.

In the first UK waterborne outbreak of cryptosporidiosis, which occurred in Ayrshire in April 1988, the potential of Cryptosporidium spp. oocysts to be waterborne was recognised by the individuals involved in the investigation. Had this not been so, the cause of the outbreak may never have been established.

Acknowledgements

The author wishes to acknowledge the work of the Badenoch Committee and is grateful to the Director of Water, Strathclyde Regional Water for permission to prepare and present this paper.

References

1. Department of the Environment and Department of Health (1990). *Cryptosporidium in water supplies*, HMSO, London.
2. Casemore, D.P. (1987) 'Cryptosporidiosis'. *Public Health Laboratory Service Microbiology Digest* 44, pp 1-5.
3. Nichols, G., and Thom, B.T. (1985) 'Food poisoning caused by Cryptosporidium: a load of tripe'. *Communicable Disease Report*, 85/17, p 3
4. D'Antanio, R.G., Winn, R.E., Taylor, J.P., Gustafson, T.L., Current W.E., Rhodes, M.M., Gary, G.W., and Zajac, R.A., (1985) 'A waterborne outbreak of cryptosporidiosis in normal hosts'. *Annals of Internal Medicine*, 103, pp 886-888.
5. Logsdon, G.S. (1987) 'Evaluating treatment plants for particulate contaminant removal'. *Journal of the American Works Association*, 79, pp 82-92.
6. Hayes, E.B., Matte, T.D., O'Brien, T.R., McKinley, T.W., Logsdon, G.S., Rose, J.B., Ungar, B.L.P., Word, D.M., Pinsky, P.F., Cummings, M.L., Wilson, M.A., Long, E.G., Hurwity, E.S. and Juranek, D.D. (1989) 'Large Community outbreak of cryptosporidiosis due to contamination of a filtered public water supply'. *New England Journal of Medicine*, 320, pp 1372 - 1376.
7. Craun, G., (1990), personal communication.
8. Smith, H.V., Patterson, W.J., Hardie, R., Greene, L.A., Benton, C., Tulloch, W.J., Sharp, J.C.M. and Forbes, G.I. (1989). 'A waterborne outbreak of cryptosporidiosis caused by post treatment contamination'. *Epidemiology and Infection*, 103; pp 703-715.
9. Benton, C. (1990) 'Experience of a waterborne outbreak of cryptosporidiosis'. *Proceedings of the First National Workshop* (Smith, H.V., and Carrington, E.G., eds) WRc publication, Medmenham, Bucks. (in press).

10. Dick, T.A. (1989) Report of an enquiry into water supplies in Oxford and Swindon following an outbreak of cryptosporidiosis during February-March 1989. Reading; Thames Water Utilities.

6. Cyanobacterial (blue-green algal) toxins and their significance in United Kingdom and European waters

L.A. Lawton and G.A. Codd

Abstract

The cyanobacteria (blue-green algae) which commonly produce mass developments in fresh and saline waters are capable of producing toxins. These compounds have been responsible for the deaths of animals, birds and fish in many countries, and have been linked with several forms of human illness through skin contact and ingestion. Evidence has accumulated over the past decade that cyanobacterial toxins occur commonly in European waters which contain cyanobacterial mass growths. This chapter summarises the recent published occurrence and types of cyanobacterial toxins found in European waters and refers to poisoning incidents. Cyanobacterial blooms were particularly abundant in the warm dry summer of 1989 in the United Kingdom. The deaths of sheep and dogs and cases of human illness ascribed to cyanobacterial toxins were followed by a more intensive and wider investigation into the extent of cyanobacterial toxicity in British waters than hitherto. Of the cyanobacterial blooms sampled from over 90 freshwaters, approximately two-thirds were toxic (lethal) according to mouse bioassay. Short term and long term needs for the recognition, quantification and management of problems which can be caused by potentially toxigenic cyanobacteria in water-bodies used for recreation, aquaculture and potable supply are briefly discussed.

Introduction

The first published account clearly implicating cyanobacteria in animal poisonings originated from Australia, where a government analyst, George Francis, attributed the deaths of sheep, cattle and horses at Lake Alexandrina, Adelaide, to drinking water containing the brackish water cyanobacterium *Nodularia spumigena*.[1,2] The perceptive report and inferences of Francis are consistent with recent investigations into livestock poisonings[3] and the presence of a hepatotoxic pentapeptide in blooms and isolates of *N. spumigena*.[4,5] In commissioning Francis to investigate the poisonings over a century ago, the Chief Inspector of Sheep to the Commissioner of Crown Lands pointed out that 'It (was) desirable that this analysis should be made to allay the feeling of doubt and uneasiness as to the cause of deaths amongst the cattle, horses and sheep near Lake Alexandrina'.[6] These sentiments have been encountered in the recent past in the UK and elsewhere in Europe among water industry workers, veterinarians, environmental health officers and others, following poisoning episodes suspected to be due to cyanobacterial toxins.

Cyanobacterial toxins are recognised to have caused the deaths of wild animals, farm livestock, pets, fish and birds in many countries and are suspected as having caused human illness in Australia, North America and Europe.[4,5,7,8,9,10] Many of the descriptions of cyanobacterial toxin poisoning incidents are incomplete, post-event accounts and awareness of the effects of cyanobacterial toxins is, despite a substantial recent increase in the literature, still lacking. These problems contribute to the lack of understanding of the incidence of the toxins in waterbodies. From a modest recent increase in the scale of cyanobacterial bloom toxicity assessment in Scandinavia, Wisconsin USA and the UK, it is clear that the occurrence of cyanobacterial toxins in blooms is far greater than would be inferred from accounts of poisoning incidents alone.[5,11] Since our understanding of the biological and toxicological effects of these toxins is still at an early stage, it is important that the occurrence and actions of cyanobacterial toxins are more thoroughly investigated in waters which are subjected to human influence and used for mans' requirements.

Properties of cyanobacterial toxins

Toxin purification and characterization from natural blooms and laboratory isolates is a rapidly developing field. It is facilitated by developments in high-performance liquid chromatography, stereochemical analysis, nuclear magnetic resonance spectroscopy and electrospray mass spectroscopy. Cyanobacterial toxins which cause death in animals, birds and fish include the hepatotoxic heptapeptide microcystins and the pentapeptide nodularin, and several neurotoxins including alkaloids and organophosphorus compounds. Comprehensive discussion of their structures and actions is beyond the scope of this chapter and the following sources are recommended for references.[4,5,7,12,13,14]. Cyanobacteria, as Gram-negative prokaryotes, also commonly produce lipopolysaccharide (LPS) endotoxins[15,16] although these appear to be only about 10 per cent as toxic as *Salmonella* LPS. 'New' cyanobacterial toxins are emerging as poisoning incidents are investigated in conjunction with cyanobacterial strain isolation and toxin purification. A notable example may be a potent hepatotoxin(s), distinct from the microcystins or nodularin, which is produced by a laboratory isolate of *Cylindrospermopsis raciborskii*.[17] This cyanobacterium dominated the bloom which was present in the drinking water supply of the human population which was severely affected by hepato-enteritis on Palm Island, Queensland.[18]

In addition to the study of molecular structure, the modes of action of cyanobacterial toxins at the molecular level are important in helping to assess their biological significance and the hazards which they may constitute to animal and human health. For example, we have recently shown that the toxic heptapeptide microcystin-LR, purified from *Microcystis aeruginosa*, is a potent and specific inhibitor of protein phosphatases PP1 and PP2A in vitro.[19] These phosphatases are major factors in the control of several cellular processes in animals, including carbohydrate metabolism, muscle contraction and cell division. It is of considerable interest that microcystin-LR inhibits PP1 and PP2A, whether from animals, plants or protists with the same specificity and about the same potency as okadaic acid (Table 1).

Table 1   *The inhibition of protein phosphatases [a] PP1 and PP2A by okadaic acid and microcystin-LR[19]*

Phosphatase	Toxin	IC_{50} [b]
PP1	okadaic acid	ca. 10.0 nM
	microcystin-LR	0.1 nM
PP2A	okadaic acid	ca. 0.1 nM
	microcystin-LR	0.1 nM

a.   Catalytic subunits purified to homogeneity from rabbit skeletal muscle.
b.   Concentration of toxin causing 50 per cent inhibition of enzyme activity.

The fatty acid polyketal okadaic acid, which is produced by marine dinoflagellates, which accumulates in mussels and is an agent of diarrhetic shellfish poisoning, is a potent tumour promoter.[20] It appears that okadaic acid and the cyanobacterial toxin microcystin-LR inhibit the protein phosphatases by binding at the same sites on the enzymes.[19] Several of the toxins of marine cyanobacteria are tumour promoters[20] and the similar molecular modes of action of microcystin-LR and okadaic acid leads unavoidably to the suspicion that microcystins may also be tumour promoters. Whether microcystins can specifically cause or promote tumours in subjects coming into contact with or ingesting the toxins, requires investigation. Indications of increased tumour numbers and growth in mice following the oral consumption of microcystin containing extracts of *Microcystis* have been obtained[9,10,21] and more research on the hazards of long term exposure to toxic cyanobacterial blooms and to microcystins in particular, is needed.

Occurrence of toxic blooms and cyanobacterial toxins in European waters

According to published accounts, toxic cyanobacterial blooms had been found in freshwaters in 16 European countries up to and including 1989. These findings are based on reports of intoxications and on laboratory toxicity assessments and refer to Czechoslovakia, Denmark, Finland, France, both former German republics, Greece, Hungary, Italy, the Netherlands, Norway, Poland, Portugal, Russia, Sweden and the United Kingdom.[13] Awareness of the incidence of cyanobacterial toxins in any particular European country, appears to be influenced by the recognition of animal poisonings and human health complaints linked to cyanobacteria, and whether (as a consequence) researchers have specifically gone out to look for toxic blooms. Understanding of the distribution of toxic cyanobacterial blooms in Europe is thus very uneven, incomplete, and in need of more systematic study. The only available data based on large-scale surveys are shown in Table 2.

Table 2 Incidence of toxicity of cyanobacterial blooms in European waters

Country	No. sites with blooms tested	No. sites with blooms toxic	Per cent Incidence of toxicity
A. Before 1989[a]			
Norway, Sweden plus Finland	51	30	59
Sweden	27	15	56
Finland	188	83	44
United Kingdom	24	18	75
B. 1989[b]			
United Kingdom	91	62	68

a. From Codd et al. (1989a)[13] and Sivonen et al. (1990).[22]
b. National Rivers Authority (1990);[24] Codd, G.A. and Beattie, K.A. (unpublished).

Toxicity assessments of cyanobacterial blooms from almost 300 sites, made between 1981 and 1989 by intraperitoneal mouse bioassay, indicated an incidence of toxicity ranging from about 45 to 75 per cent. The toxic blooms commonly consisted of species of *Microcystis, Anabaena, Oscillatoria* and *Aphanizomenon* and less often consisted of *Gomphosphaeria, Coelosphaerium, Nodularia, Nostoc* and *Cylindrospermum* spp. The results shown in Table 2, whilst indicating that the likelihood of encountering toxicity in a cyanobacterial bloom is high, belie a highly dynamic complex of changes in toxicity per unit cyanobacterial biomass at individual lakes within a single bloom season.[23,24] The causes of variations in toxin levels per unit bloom biomass versus time and in the horizontal and vertical dimensions in waterbodies, and of changes in the relative levels of individual toxins in a mixture, for example of microcystin variants,[25] are not understood.

Our investigations into cyanobacterial bloom toxicity throughout the eighties[8] indicated that the incidence of toxicity in UK freshwaters was probably about the same as elsewhere in Europe, and in Australasia and the Americas where toxic blooms have been more often recognised in poisoning events. With the exceptions of some fish kills, isolated cattle deaths and human skin irritations attributed to toxic cyanobacterial blooms in the UK,[13] there was little recent evidence prior to 1989 that cyanobacterial toxins presented a health hazard to animals and human users of British waters. This position changed rapidly in September 1989.

Events in 1989

United Kingdom

A high incidence of cyanobacterial blooms was encountered in British freshwaters during the long hot summer and mild autumn of 1989, including blooms of *Anabaena* and *Aphanizomenon*, followed by *Microcystis* at Rutland Water in Leicestershire, England. This 1,260 hectare lake, possibly the biggest man-made waterbody in Western Europe, is part of a supply network for the provision of water to 1.5 million people. *Microcystis aeruginosa* scum accumulated to a depth of several centimetres and a width of one to about fifty metres around the entire shore-line of the lake. Rutland Water is a highly attractive recreational facility with good access for the public for activities such as water sports, fishing, bird watching, picnics. It is also a Site of Special Scientific Interest. Much of the shore-line also gives access to sheep pasture. Twenty sheep and fifteen dogs were reported to have died at Rutland Water between August 21 and mid-September after alleged contact with the scum. Samples of *M.aeruginosa* scum received at this laboratory in early September were lethal (hepatotoxic) by mouse bioassay and the 994 Daltons heptapeptide microcystin-LR was characterized by mass spectrometry as the principal cyanobacterial toxin present. These observations, together with the signs of illness in the sheep and dogs, times of death, the pathological and biochemical analyses by veterinarians and our findings of microcystin-LR in the rumen contents of the sheep led this laboratory, and a National Rivers Authority (NRA) report,[24] to suspect strongly that the animals died from microcystin poisoning.

The animal deaths at Rutland Water were followed in September by illnesses among army junior soldiers who had been canoeing and swimming at Rudyard Lake, Staffordshire, England. Rudyard Lake at the time also contained an intense bloom of *M.aeruginosa* from which we again isolated microcystin-LR. The canoeing exercises included 360° rolls and swimming exercises occurred through the cyanobacterial scum . Eight soldiers subsequently reported influenza-like symptoms and two became seriously ill and were admitted to hospital. Their symptoms occurred the day after being in contact with the scum and included sore throat, blistering in and around the mouth, abdominal pain, vomiting, diarrhoea, fever and pleuritic pain on the left side. One had difficulty in walking and was periodically confused with hallucinations for two days.[26] Both patients displayed basal consolidation of the left lung, with signs of an atypical pneumonia which persisted for about 4 days. The soldiers were well enough for discharge after one week. Bacteriological and viral tests for known waterborne pathogens were negative. One of the soldiers displayed elevated liver enzyme levels in the blood stream which are consistent with signs of microcystin poisoning in laboratory animals. Both showed low platelet counts on admission to hospital, which subsequently recovered. These observations in total have led the investigators to infer that the soldiers' illnesses were caused by contact with, and ingestion, and perhaps inhalation, of microcystin-containing *Microcystis* scum.[26] Reports were also received of skin irritations, nausea and vomiting among several sail-boarders who had been in contact with *Microcystis* scum at Rutland Water and other lakes with toxic cyanobacterial blooms in September 1989.

The events at Rutland Water and Rudyard Lake coincided with the recent restructuring of the water industry in England and Wales, including the establishment of the NRA as the national body with responsibilities and powers relating to the monitoring, maintenance and, where necessary, improvement of water quality, including surface water intended for the abstraction of drinking water. An immediate response by the NRA to the poisoning events occurred, with actions designed to obtain a national overview of the incidence of toxic cyanobacterial blooms, in addition to more intensive investigations with measures taken at the sites where incidents linked to toxic cyanobacteria had already occurred. Such actions included the monitoring of freshwaters for cyanobacterial species and abundance, and toxicity assessment; informing owners and appropriate government officers in environmental/public health and in agriculture/fisheries of the findings; suggesting precautions; issuing press releases and responding to the large and prolonged interest of the media and of the general public.

Public interest in the poisoning events and in the obvious magnitude of the cyanobacterial scums at several sites was very high. Notices were posted by owners at several affected lakes and reservoirs which remained in place for several weeks. The impact of the animal poisonings and human illnesses and of the warning to avoid contact with cyanobacterial scum, particularly by children, to prevent animals from drinking water containing scum, and of the high public and media interest, had major effects on water-based activities for the remainder of 1989. Activities affected included cancellations of water sports events and fishing competitions, with a major decline in the number of people visiting and using freshwaters for recreational purposes, with resulting associated economic losses.

The NRA monitoring exercise throughout England and Wales included inspections at 915 different waters in 1989. Of these, 686 were sampled for phytoplankton and 594 of these had cyanobacteria as the dominant group. In England and Wales 169 waters were considered by the NRA to present cyanobacterial-related problems. Toxicity assessments of cyanobacterial bloom or scum samples from 78 sites in the NRA programme were made, largely by this laboratory working with the NRA. Of these, 53 (68 per cent) were positive by mouse bioassay. These toxicity data are included in the overall estimate for cyanobacterial bloom toxicity of which we are aware for 1989 (Table 2), the balance being accounted for by cyanobacterial bloom toxicity testing by this laboratory from other sites in the UK, particularly Scotland. The total estimate of the incidence of cyanobacterial bloom toxicity (68 per cent) for 1989 was broadly in agreement with values obtained in previous years (Table 2). Together with external specialists in limnology, environmental and cyanobacterial toxicology the NRA established a 'toxic algae task group' in 1989 which has continued to collect data and to identify, formulate and recommend measures for the short and long term management and control of cyanobacterial toxin-related problems, as encompassed by the responsibilities of the NRA. The findings of the 1989 survey, with details of steps taken and future recommendations, have been published recently.[24]

Elsewhere in Europe

Animal deaths and human health problems attributed to cyanobacterial toxins in Scandinavian countries have been documented for many years and the level of public awareness of the hazards presented by cyanobacterial blooms in these countries is relatively high in Europe. This may contibute to the more recent decrease in reports of poisoning incidents, as blooms are avoided by water-users and farmers watering livestock. However, isolated reports communicated to the author by O.M. Skulberg, T. Willen, T. Lindholm and V. Tchernajenko for 1989 have included:

Norway

Bird kills occurred at Lake Ostensjo, Oslo, which contained a toxic bloom of *Microcystis* and *Anabaena*. Hepatotoxic *Microcystis* blooms occurred at Lake Akersvatn, Oslo, at several lakes near Stavanger and a neurotoxic bloom of *Aphanizomenon* and *Anabaena* was present in the Haldenvass Draget water course. Animal deaths were not reported but several lakes were closed to the public.

Denmark

Microcystis blooms were common in several lakes but no animal deaths were reported. A high level of awareness exists in Denmark through public warning notices and leaflets following the deaths of dogs and cattle in recent years.

Sweden

High media interest centred on a highly toxic bloom of *Anabaena* at Lake Tullingesjon near Stockholm. Bathing was forbidden and no animal poisonings occurred. Toxic *Microcystis* and *Anabaena* blooms and closures also occurred in northern and southern Sweden, including at Lake Finjasjon a drinking water supply where cases of skin irritation, vomiting and diarrhoea had occurred in 1987.

Finland

Hepatotoxic blooms were present in Lake Ostra Kyrksundent, Aland and at Dragsfjord, southern Finland, where complaints occurred of cyanobacterial off flavours in the drinking water, which also contained free microcystin and *Oscillatoria* filaments.

USSR

Hepatotoxic *Microcystis* blooms were recorded in Lake Ladoga (Leningrad), at Kiev, and in the Pusch Gulf where fish kills in *Microcystis* and *Aphanizomenon* blooms also occurred.

Monitoring, control and management

In the aftermath of the 1989 poisonings in the UK, a considerable effort is being directed by the NRA and the water body owners towards the future monitoring of cyanobacterial blooms and toxins. A trial monitoring system has been operated by the NRA throughout the summer and autumn of 1990 throughout its ten regions in England and Wales. This has centred on the identification and approximate quantification of the species of cyanobacteria which are proven and potential toxin-formers. Inspections and sampling protocols have been applied in terms of the demands made of individual waters, which have been categorised in the first instance according to the risk of contact with and ingestion of cyanobacterial scum by humans, livestock and pets. A public record, established by the NRA, has regularly listed freshwaters where potentially toxic cyanobacterial blooms have occurred and appropriate warnings have been given to water owners and operators, environmental health and agriculture officials. This procedure will be reviewed and modified as necessary to integrate with procedures under consideration for the monitoring of the toxins and the removal and control of cyanobacterial blooms. Public awareness of the hazards presented by toxic cyanobacterial blooms has been increased considerably in England and Wales by the production and wide-scale dissemination of a warning leaflet in 1990.

Requirements for the control and management of toxic cyanobacterial blooms have been identified.[13,24,27] Physical controls for consideration include destratification, light exclusion for small facilities and scum coralling. Nutrient deprivation, biological control and the use of algicides (though not if they lyse cyanobacteria and cause the release of toxins into the surrounding water) are also options which require considerable research and development. The strategy of choice will eventually depend on the physical characteristics of the waterbody and the use to which it is subjected. The common occurrence of toxic cyanobacterial blooms in waterbodies which are intensively used for access, leisure and drinking in the UK and Europe, and the hazards which their toxins present to human and animal health, also combine to emphasise the need for accredited capabilities for the recognition and quantification of the toxins in waterbodies and supplies. Furthermore, although some procedures for the limitation or removal of cyanobacterial blooms have been practised on site, or investigated at laboratory-scale for some years, it is now apparent that control options should also be specifically evaluated for their effects on the distribution and levels of the toxins themselves.

Acknowledgements

It is a pleasure to thank the following for their kind help in obtaining information about the investigations of George Francis into suspected cyanobacterial poisonings of over a century ago: Alison J. Hoyle, the Public Record Office, North Adelaide, Mrs B. Mayfield, The Matlock Library of South Australia, Adelaide and Lance Schlipalius, Betatene Ltd, Cheltenham, Victoria, Australia.

References

1. Francis, G. (1878). *Nature* (London) 18, pp 11-12.
2. Valentine, C.J. (1878a). *Report of the South Australian Government Gazette*, Feb. 7., pp 278-280.
3. Main, D.C., Berry, P.H., Peet, R.L. and Robertson, J.P. (1977). *Australian Veterinary Journal*, 53, pp 578-581.
4. Carmichael, W.W. (1989). In: *Natural Toxins: Characterization, Pharmacology and Therapeutics*. C.L. Ownby and G.V. Odell (Eds), Pergamon Press, Oxford, pp.3-16.
5. Codd, G.A., Bell, S.G. and Brooks, W.P. (1989a). *Water Science and Technology* 21, pp 1-13.
6. Valentine, C.J. (1878b). Letter to the Secretary, Crown Lands from the Inspector of Sheep Office, Adelaide, July 10, 1878.
7. Carmichael, W.W. (1986). In: *Advances in Botanical Research*. J.A. Callow (Ed.), Vol. 12, Academic Press, London, pp 47-101.
8. Codd, G.A. and Poon, G.K. (1988). In: *Biochemistry of the Algae and Cyanobacteria*. L.J. Rogers and J.R. Gallon (Eds), Clarendon Press, Oxford, pp. 283-296.
9. Falconer, I.R. (1989), *Toxicity Assessment*, 4, pp 175-184.
10. Falconer, I.R. (1990). *Environmental Toxicology and Water Quality* (in press).
11. Repavich, W.M., Sonzogni, W.C., Standridge, J.H., Wedepohl, R.E. and Meisner, L.F. (1990). *Water Research*, 24, pp 225-231.
12. Gorham, P.R. and Carmichael, W.W. (1988). In: *Algae and Human Affairs*, C.A. Lembi and J.R. Waaland (Eds). Cambridge University Press, Cambridge, pp. 403-431.

13. Codd, G.A., Brooks, W.P., Lawton, L.A. and Beattie, K.A. (1989a). In: *Watershed '89, The Future for Water Quality in Europe,* D. Wheeler, M.J. Richardson and J. Bridges (Eds), Vol. II, Pergamon Press, Oxford, pp 211-220.
14. Lawton, L.A., Hawser, S.P., Jamel Al-Layl, K., Beattie, K.A., Mackintosh, C. and Codd, G.A. (1990). In: *Proceedings of the Second Biennial Water Quality Symposium: Microbiological Aspects,* G. Castillo, V. Campos and L. Herrera (Eds), University of Chile, Santiago, pp 83-89.
15. Raziuddin, S., Siegelman, H.W. and Tornabene, T.G. (1983). *European Journal of Biochemistry,* 137, pp 333-336.
16. Martin, C., Codd, G.A., Siegelman, H.W. and Weckesser, J. (1989). *Archives of Microbiology,* 152, pp 90-94.
17. Hawkins, P.R., Runnegar, M.T.C., Jackson, A.R.B. and Falconer, I.R. (1985). *Applied and Environmental Microbiology,* 50, pp 1292-1295.
18. Bourke, A.T.C., Hawes, A.B., Nielson, A. and Stallman, (1983). *Toxicon Supplement,* 3, pp 45-48.
19. Mackintosh, C., Beattie, K.A., Klumpps, S., Cohen, P. and Codd, G.A. (1990). FEBS letters, 264, pp 187-192.
20. Fujiki, H., Suganuma, M. and Sugimura, T. (1989). Environmental and Carcinogenicity Reviews, *(Journal of Environmental Science and Health)* C7, pp 1-51.
21. Falconer, I.R. and Buckley, T. (1989). *Medical Journal of Australia,* 150, p 351.
22. Sivonen, K., Niemela, S.I., Niemi, R.M., Lepisto, L., Luoma, T.H. and Rasanen, L.A. (1990). *Hydrobiologia ,* 190, pp 267-275.
23. Codd, G.A. and Bell, S.G. (1985). *Journal of Water Pollution Control,* 34, pp 225-232.
24. National Rivers Authority (1990). 'Toxic Blue-Green Algae'. *Water Quality Series No.2 ,* National Rivers Authority, London, 128 pp.
25. Kaya, K and Watanabe, M.M. (1990). *Journal of Applied Phycology,* 2, pp 173-178.
26. Turner, P.C., Gammie, A.J., Hollinrake, K. and Codd, G.A. (1990). *British Medical Journal,* 300, pp1440-1441.
27. Reynolds, C.S. (1987). In: *Advances in Botanical Research,* J.A. Callow (Ed.), Vol. 13, Academic Press, London, pp. 67-143.

7. New treatments for pesticides and chlorinated organics control

D.M. Foster, A.J. Rachwal and S.L. White

Abstract

Sources of pesticides and chlorinated organic compounds potentially found in water sources are reviewed. International and UK drinking water standards for these compounds are discussed, and compared with concentrations found in ground and surface waters derived from lowland agricultural and urban areas.

Water soluble herbicides including triazines, phenylamides (urons) and chlorophenoxy acids are the most commonly reported types of pesticide compounds found above 0.1ug/l in water sources. The more toxic, and less water soluble, organochlorine and organophosphorus pesticides are rarely reported.

Chlorinated organic solvents are reported as contaminants of some groundwaters. Disinfection by-products, in particular trihalomethanes, formed during chlorination of waters containing natural organics, are widely reported, sometimes at concentrations above current UK standards.

The effects of conventional water treatment processes on the concentrations of these organic compounds are discussed. Additional water treatment processes designed specifically for trace organic micropollutant removal are identified and reviewed. Ozonation, activated carbon adsorption and air stripping are acknowledged as appropriate advanced water treatment process solutions. Developments in novel technologies include organics destruction via advanced oxidation processes, biological processes, membrane technology and novel adsorbents.

Thames Water plc. is contributing to research in this area, investing over £5 million in pilot and large-scale trials, prior to a £200-£300 million capital investment programme in advanced water treatment in the 1990s.

Introduction

In recent years drinking water quality has become a major issue for public and political debate. Water quality issues in the public eye include: nitrates, lead, aluminium, trihalomethanes (THMs) and pesticides. In recent months concern over pesticides has come to the fore with the publication of the BMA report[1] 'Pesticides, Chemicals and Health' and the Friends of the Earth's legal action against the Secretary of State for accepting pesticide 'undertakings' from water suppliers.

Table 1   Pesticides reported in water supplies in England and Wales at concentrations above 0.1ug/l during 1988[4]

Pesticide	Main type / use
Atrazine	H - non agricultural
Carbetamide	H - various crops
Chlortoluron	H - cereals
Clopyralid	H - various crops and non-ag
2, 4-D	H - various crops and non-ag
Dicamba	H - various crops and non-ag
Dichlorprop	H - cereals
Dimethoate	I - cereals
EPTC	H - pre-emergence
Isoproturon	H - cereals
Linuron	H - various crops
Malathion	I - various crops
MCPA	H - cereals and non-ag
MCPB	H - various crops
Mecoprop	H - cereals and non-ag
Prometryne	H - various crops
Propazine	H - no approved UK uses (not confirmed)
Propyzamide	H - various crops
Simazine	H - non agricultural
Triallate	H - cereals

Key to Table 1   H - Herbicide
                         I - Insecticide
                         non-ag - non-agricultural use

The pesticide 'problem'

In a regulatory sense, pesticides are a major problem to water suppliers. The UK Standard[2] for individual pesticides in drinking water is 0.1ug/l, derived from the 1980 EC Drinking Water Directive, parameter 55.[3] Table 1 lists pesticides which were reported in water supplies in England and Wales at a concentration above 0.1ug/l on a least one occasion during 1988.[4] Nearly all are herbicides.

The presence of pesticides above the standard in drinking water means that, technically speaking, such water is not 'wholesome'. Because of this, water suppliers have had to give undertakings to the Secretary of State detailing the steps they are going to take to secure or facilitate compliance. These steps are involving water suppliers in major research and investment programmes on treatment methods to remove pesticides.

The origins and validity of the pesticide standard have been discussed at length elsewhere and will not be reported here. The key point is that the European (and hence the UK) standard applies to all pesticides irrespective of their toxicity. This puts the standard at variance with all other limits for pesticides in drinking water which have been developed throughout the world (Table 2).

In every other case, regulators have established limits based on toxicological data for individual substances. Although the various countries have developed different limits, most are considerably greater than 0.1ug/l.

Sources of pesticides

While applications to agricultural crops undoubtedly account for some of the pesticides found in water supplies, eg. isoproturon used on cereals, they are not the whole story.

Atrazine and simazine, the two most commonly reported pesticides in drinking water, have extremely limited use in agriculture but are widely used to control weeds in non-agricultural situations such as on road verges, railways and footpaths. This might suggest that they are particularly polluting chemicals, since the area of application is relatively small. In fact the appearance of atrazine and simazine in water sources is readily explained by the very high application rates in non-agricultural situations, and the fact that they are applied to hard surfaces with fast routes to surface and groundwaters via surface drains and soakaways.

Table 2   *Comparison of standards and guidelines for pesticides found in drinking water sources*

	Standard / Guideline / Guide Level				
Pesticide	EC	WHO	US	Canada	UK(DoE) (a)
Atrazine	0.1	2	3 (b)	60 (c)	2
Bentazone	0.1	25	20 (e)	-	-
Chlortoluron	0.1	-	-	-	80
2, 4-D	0.1	-	70 (d)	100	1,000
Glyphosate	0.1	-	700	280	1,000
Isoproturon	0.1	-	-	-	4
Lindane	0.1	-	4	4	3
MCPA	0.1	0.5	4 (e)	-	0.5
Mecoprop	0.1	-	-	-	10
Simazine	0.1	17	1 (b)	10 (c)	10

Key to Table 2
(a)   DoE - advisory values.[4]
(b)   Tentative Standard, under development.
(c)   Interim maximum acceptable concentration.
(d)   Proposed Standard, under development.
(e)   Lifetime health advisory, not a regulatory standard.

It is noteworthy to calculate that some 2kg of pesticide reaching the total daily UK public water supply would exceed the 0.1ug/l standard. Thousands of tons of herbicides are used annually.

Input controls for pesticide reduction

The manufacturers, users and regulators of pesticides are increasingly aware that perfectly legal and carefully controlled uses of pesticides may still give rise to levels of pesticides in water sources above the 0.1ug/l limit. With the maxims that the 'polluter pays' and 'prevention is better than cure' in mind, all parties are looking at ways of controlling inputs of pesticides to the environment. For example, Thames Water has recently asked major identified users to review and reduce their use of pesticides, and where possible to move to substances that are less likely to reach source waters.

Sources and standards for other chlorinated organic compounds

Other generic groups of organics, such as industrial solvents and trihalomethanes (THMs) may, on a local scale, be as problematic as pesticides.

Problems with industrial solvents, such as tetrachloroethene (also commonly referred to as perchloroethene (PCE)) and trichloroethene (TCE,) are typically confined to groundwaters and may result from deliberate dumping, spillages or poorly constructed landfill sites. In contrast, THMs typically arise in drinking water as a result of water treatment processes where chlorine reacts with naturally occurring organic matter present in raw waters. THMs are usually only a feature of surface water.

In contrast with the pesticides situation, there are no standards for solvents and THMs in the European Drinking Water Directive, although there is a non-statutory guide level of 1 ug/l.

Without the constraints of a EC limit, the UK standards[2] for solvents and THMs have been derived from toxicological data. While this means that UK standards accord well with limits established by other regulatory agencies, (see Table 3), it does mean that exceedances of standards indicate some, if only theoretical, health risks.

Table 3 shows a considerable degree of uniformity in standards for solvents and THMs. However, there are limits, current and proposed, that would be more difficult for UK suppliers to achieve. For example in the US the National Research Council has concluded that the 100 ug/l maximum contaminant level should be reduced. Limits of 25 or 50 ug/l have been suggested. In Germany the Federal Health Office has recommended a limit value for total THMs of 25 ug/l (as an annual average). In the Netherlands, a level of >1ug/l acts as a trigger level to institute a full sampling programme.

Pesticide and solvent levels in source waters

Occurrence of pesticides in source waters

A survey[5] of UK surface and groundwater sources identified the triazine herbicides, atrazine and simazine as the most common pesticides in both ground and surface waters. Concentrations of up to 0.6ug/l for each were detected frequently in surface waters with maximum concentration values of 9.0ug/l and 2.8ug/l reported for atrazine and simazine respectively. Other pesticides

Table 3 Standards / guidelines for solvents and trihalomethanes

Substances	Standards / Guidelines (ug/l)				
	UK	WHO	US	Canada	Australia
Tetrachloromethane (carbon tetrachloride)	3 (a)	3	5	5	3
Tetrachloroethene (perchloroethene (PCE)	10 (a)	10	5 (b)	- (c)	10
Trichloroethene (TCE)	30 (a)	30	5	- (c)	30
Trihalomethanes (sum of 4 compounds)	100 (d)	- (e)	100	350 (c)	200

Key to Table 3

(a) Limit as an annual average.
(b) Proposed Standard, under development.
(c) Under review.
(d) Limit as a three monthly average.
(e) 30 for chloroform.

frequently found[5,6] in surface waters included the chlorphenoxy acid herbicide mecoprop, the phenylamide herbicide isoproturon and the organochlorine insecticide lindane.

The seasonal pattern of pesticide use, together with variations in rainfall, run-off and river flows, will lead to considerable variations in pesticide concentration in surface waters. Thames Water has monitored the concentrations of approximately 40 individual pesticides at three sites on the Rivers Thames and Lee over a 15 month period. Nine pesticides have been identified as occurring at concentrations above 0.02ug/l on a regular basis. Eight are herbicides, whilst the other is the organochlorine insecticide lindane. The results of the survey are summarised in Table 4.

Table 4 Identity, concentration range and frequency of occurrence of pesticides detected in the Rivers Thames and Lee, July 1989-September 1990

Pesticide	Concentration		Occurrence
	Range of means ug/l (a)	Maximum ug/l	per cent of samples >0.02 ug/l
Atrazine	0.27 - 0.47	1.60	100
Simazine	0.13 - 0.37	1.80	100
Isoproturon	0.14 - 0.22	1.44	92
Diuron	0.11 - 0.23	0.9	89
Chlortoluron	0.05 - 0.06	0.23	61
Mecoprop	0.07 - 0.09	0.18	100
MCPA	<0.02 -<0.02	0.06	37
2, 4-D	0.02 - 0.06	0.30	90
Lindane	0.02 - 0.03	0.08	94

(a) Range of mean concentrations at 3 sites (12-15 monthly samples)

It should be noted that the maximum detected concentrations are all less than the health related guide levels given by the DoE (ref. Table 2), and that, with the exception of lindane, the more toxic organochlorine and organophosphorus pesticides were not present.

Monitoring at groundwater sites in the Thames Water area has confirmed other investigators' findings that atrazine and simazine are present in many sources, particularly in urban areas where highway and railway drainage can discharge to soakaways. Concentrations have been generally similar to those found in the surface waters.

Seasonal variation in pesticide levels in surface water sources is illustrated in Figures 1, 2 and 3, which show results for one sample point on the River Thames. The peak atrazine, simazine and diuron concentrations were recorded in the summer period. It is not known whether this is a consequence of higher application rates or lower river flows. Highest concentrations of isoproturon, chlortoluron and mecoprop occurred in the winter and spring. This would be predicted, as the predominant use for these compounds is as pre-emergent herbicides for winter and spring sown cereal crops.

New Health Considerations in Water Treatment

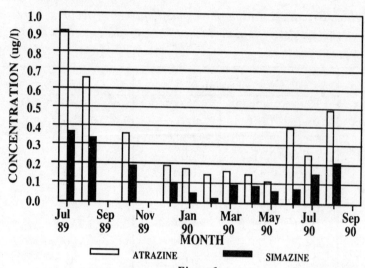

Figure 1
Triazine Herbicide Concentrations in the River Thames July 1989 - September 1990

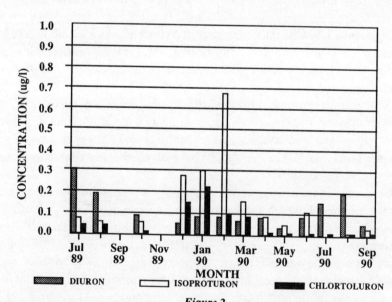

Figure 2
Phenylamide herbicide concentrations in the River Thames July 1989 - September 1990

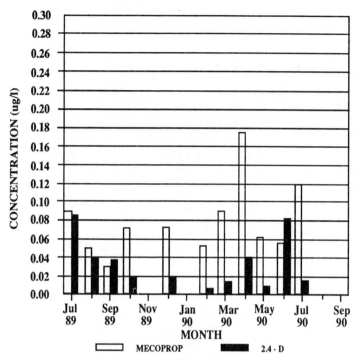

Figure 3
Chlorphenoxy Acid Herbicide Concentrations in the River Thames July 1989 - September 1990

Occurrence of chlorinated organic solvents in groundwaters

The most commonly found contaminants are those compounds for which standards have been set in the Regulations:[2] carbon tetrachloride, TCE and PCE. High levels of other solvents have been reported at some locations in the UK, including dichloromethane, di- and tri- chloroethanes and chlorobenzenes.

It is difficult to establish the full extent of the problem with volatile solvents in the UK, as past monitoring for these compounds has not been extensive,[7] and incidence and behaviour will vary widely from source to source. In some instances current pollution may be a consequence of past industrial activity, with low aquifer recharge rates meaning that continuous contamination over a long period is likely. In other cases, rapid variations in pollutant concentration may occur as solvents are transported quickly from point of discharge to the borehole location. Monitoring of groundwater at one Thames Water site showed that TCE concentration rose from 10 to 140ug/l, and then fell back, in a two month period.

Effects of conventional treatment processes

Effect of conventional treatment on pesticides

Most lowland conventional water treatment plants fall into one of three groups depending on water source and type of treatment.
1. Surface water treatment by storage, chemical coagulation, clarification, rapid sand filtration and chlorination.
2. Surface water treatment by storage, slow sand filtration and chlorination.
3. Groundwater treatment by disinfection only.
    Optional - aeration for CO_2 stripping.
    - filtration for Fe/Mn removal.

A considerable amount of academic and research literature has been published on pesticides in the environment, but little operational data on the effectiveness, or otherwise, of conventional treatment processes is available. Most studies concern higher pesticide concentrations than those usually found in drinking water, and it is not always possible to assess the effectiveness of such processes in reducing pesticide concentrations to less than 0.1ug/l.

A review[6] of published data concluded that both types of surface water treatment, chemical coagulation and slow sand filtration, were only effective in removing low solubility pesticides such as the organochlorine compounds, except where pesticides were complexed with humic material. Chlorination was reported as effective in breaking down the phenylamide herbicides and organophosphorus insecticides.

Thames Water has recently concluded a survey of pesticide levels at various process stages at a number of conventional treatment works. The treatment streams and sample points at two of these are shown in Figures 4a and 4b, and representative pesticide results are summarised in Table 5.

Reservoir storage tended to balance pesticide loads, reducing and delaying peak concentrations to further treatment.

Results from the chemical coagulation works show good removal of some of the phenylamide herbicides by the pre-chlorination and chemical clarification stage. The effect is attributed to the pre-chlorination rather than the chemical coagulation. The concentrations of the triazines, the chlorphenoxy acids and lindane were largely unchanged.

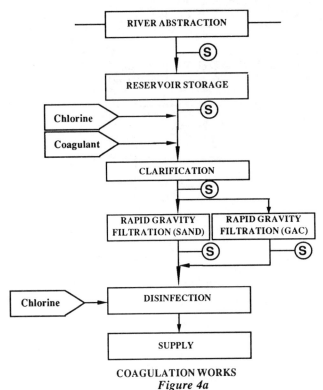

COAGULATION WORKS
Figure 4a
Treatment process schematics and location of sampling points at treatment works

Results from the slow sand filtration works show that lindane and the chlorphenoxy acids are partially removed, while the triazines and the phenylamides are largely unchanged.

Data from one Thames Water borehole source confirm that aeration and disinfection are ineffective in removing atrazine and simazine.

Effect of conventional treatment on chlorinated organic solvents

Chlorinated organic solvents are predominantly a problem at groundwater sources, where chlorination is usually the only 'conventional' treatment process. This will have no beneficial effect on solvent concentrations.

At sites where aeration for CO_2 removal is practised, transfer of volatile compounds from the liquid to the gas phase may lead to some reduction in solvent concentration.

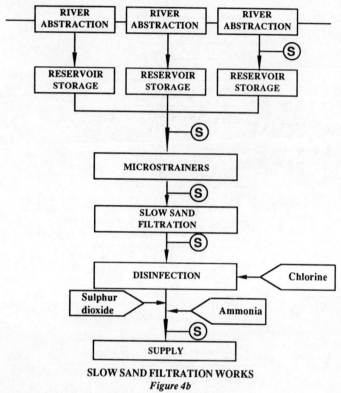

SLOW SAND FILTRATION WORKS
Figure 4b
Treatment process schematics and location of sampling points at treatment works

Trihalomethane formation in conventional treatment

There is much published data on the formation of trihalomethanes (THMs) in conventional treatment, and a detailed review is outside the scope of this chapter. It is sufficient to note that THMs are formed when water containing natural organic substances, such as humic and fulvic acids, is dosed with chlorine. The potential for formation of THMs depends on the concentrations of THM precursor compounds in the raw water, and the treatment stream used.

THM formation at chemical coagulation works

At chemical coagulation works the use of pre-chlorination before chemical dosing and clarification is a major cause of THM formation. High total organic carbon (TOC) concentrations, high chlorine doses and long contact times

*Table 5   Mean pesticide concentrations in conventional treatment,
            July 1989 - September 1990*

Treatment stage	Mean pesticide concentration (ug/l)			
	Atrazine	Iso-proturon	Meco-prop	Lindane
Coagulation works				
River water	0.27	0.18	0.07	0.016
After reservoir storage	0.25	0.15	0.05	0.012
After clarification	0.24	< 0.01	0.03	0.010
After sand filtration	0.22	< 0.01	0.03	0.010
Slow sand filtration works				
River water	0.33	0.14	0.08	0.028
After reservoir storage	0.22	0.27	0.10	0.016
After slow sand filtration	0.24	0.24	0.03	0.007
After final disinfection	0.22	0.22	0.03	0.009

Lowering the pre-chlorination dose and moving the point of application to later in the process stream will reduce the formation of THMs. This is sometimes enough to allow compliance with the 100 ug/l total THM standard, but the 30 ug/l WHO chloroform guide level may still be exceeded. Chlorination for final disinfection and the maintenane of a free chlorine residual in supply will also result in further THM formation. Cessation of pre-chlorination has caused algal and slime growth problems in filters and clarifiers.

The use of alternative pre-oxidants and disinfectants which do not form THMs, such as ozone, chlorine dioxide and preformed monochloramine, has been extensively investigated.[8,9] The use of chlorine dioxide can lead to taste and odour problems and the introduction of undesirable inorganic chlorite and chlorate ions into supply. Chloramine is a much less powerful disinfectant than free chlorine, and requires high doses and/or long contact times to match current disinfection practices. Biological formation of nitrites in the distribution system can also be a problem. The use of ozone is discussed later in this chapter.

THM formation at slow sand filtration works

The slow sand filtration process is used by Thames Water to supply the majority of London's water, some 2 million m³/day.[10] The process does not use pre-chlorination, thus preventing THM formation prior to final disinfection. Thames' practice is to follow slow sand filtration with primary disinfection, using chlorination to give a free chlorine residual of at least 0.5mg/l after 30 minutes contact. The water is then dechlorinated to 0.3mg/l before the addition of ammonia to produce a chloramine residual into supply, preventing further THM formation in the distribution system.

This practice in Thames Water minimises the formation of THMs, with total THM levels at the customer's tap being typically 40-70ug/l, which meets the current UK standard.[2] Chloroform concentrations are generally 15-35ug/l.

Options for advanced water treatment

It is clear that conventional treatment processes will be unable consistently to meet all current and proposed future standards for pesticides, THMs and chlorinated organics.

In an earlier review[11] of the literature and of existing practice in Europe, the authors and colleagues concluded that the only operationally proven advanced treatment solutions employ activated carbon for adsorption of pesticides and chlorinated solvents, combinations of ozone and activated carbon for control of trihalomethane formation potential and air stripping to remove volatile organic solvents and pre-formed THMs.

The possible options for introduction of advanced processes to conventional surface water treatment are illustrated in Figures 5 and 6.

A number of 'advanced oxidation processes' (AOPs) showed promise for complete destruction of organic micropollutants, while other novel processes were identified as possible future options.

The present status of these advanced treatment technologies, together with Thames Water's experience in their application, is reviewed in the following sections.

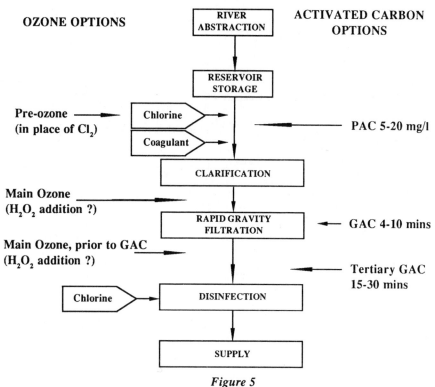

Figure 5
Conventional chemical coagulation based lowland water treatment - options for activated carbon and ozone

Ozonation processes

Ozone for pesticide removal

Many papers and reviews on the effects of ozone on pesticides in water have been published.[6,12] However, much of the reported data is for the organochlorine and organophosphorus pesticides rarely found today in water sources, or for pesticide levels and ozone doses several orders of magnitude above those found in water treatment.

The pesticides of most interest to the UK Water Industry are the triazine, phenylamide (uron) and chlorphenoxy acid herbicides. Triazines are reported as having some reactivity with ozone in water. Between zero and 90 per cent

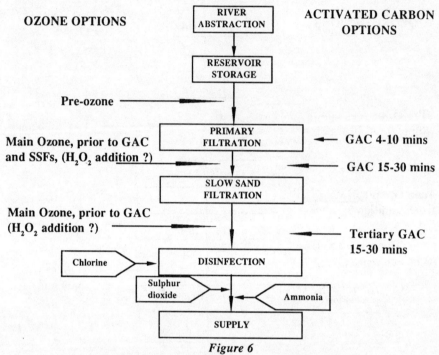

Figure 6
*Slow sand filtration based lowland water treatment -
options for activated carbon and ozone*

destruction has been achieved depending on water type, ozone dose and contact time.[6,12] Work by the authors on natural waters and flow through pilot ozone contactors indicates that 30-50 per cent removal of atrazine may be achieved with an ozone dose of 2 mg/l and 5 minute contact time. Similar results have been reported in France.[13,14]

Phenylamides such as diuron and isoproturon are reported as reactive to ozone.[12] The authors' results have shown at least 90 per cent reaction with ozone in natural waters at pilot scale.

Chlorphenoxy acids such as MCPA are reported as reacting readily with ozone.[12] The authors' results confirm this finding with almost complete removal of MCPA but reduced reactivity with mecoprop and 2,4-D.

The overall conclusion is that ozone will reduce pesticide levels, some almost completely, but that the triazines are not sufficiently reduced to eliminate the need for subsequent GAC adsorption.

Ozone for THM precursor removal

The findings of many authors[15,18,19] are that THM precursors, present in the TOC and humic and fulvic acids in water, are not totally destroyed by ozone. They are oxidised, forming aldehydes, ketones and acids which are more bio-degradable, increasing the assimilable organic carbon (AOC) and enabling removal of potential THM precursors in a subsequent biological stage.

However, if a biological process is not included after ozonation, the increased AOC levels could allow additional growth of microorganisms in the distribution system.

There is little evidence that ozone destroys preformed THMs, though a small amount of air stripping may occur.

Ozone for chlorinated solvent removal

There is little reported evidence that ozone can provide significant removal of chlorinated solvents from water. Some air stripping may occur in an ozone contactor.

Activated carbon adsorption

Activated carbon is generally produced by a controlled combustion of coal, peat, coconut shells or wood. This activation process produces a porous material with a large surface area and a high affinity for organic compounds. It is normally used either as a powder (PAC) or in a granular form (GAC). When the adsorption capacity of the carbon is exhausted, it can be regenerated thermally in a reactivation furnace. Regeneration is usually only applied for high value GAC.

Activated carbon for taste and odour control

The UK Water industry has, until recently, mainly used activated carbon for taste and odour control, utilising its capacity to remove earthy and musty tasting geosmin and 2-methyl isoborneol or synthetic micropollutants such as chlorinated phenols.

PAC doses of 5-20 mg/l or GAC empty bed contact times (EBCT) of 4-8 minutes have typically been employed for this purpose, with a carbon life of up to 3 years or 300,000 bed volumes (BV) before exhaustion. In this application GAC has generally been used as a sand replacement in existing rapid gravity filter shells.

It should be noted that many plants have experienced problems with short filter runs at high PAC doses.

Activated carbon for pesticide removal

Activated carbon is widely reported[6,17] as being capable of removing many of the pesticides found in water sources, but a number of process considerations remain to be resolved (ref. Figures 5 and 6):

* Is the use of PAC for pesticide removal a viable option?
* Which type of GAC is best for pesticide removal?
* Where in the process stream should GAC be installed?
* What EBCT should be adopted, and what will be the bed-life before regeneration is required?
* Does the use of ozone prior to GAC increase the bed-life for pesticide removal?

Pilot and full-scale trials are still the only reliable method of obtaining all the process design criteria for specific pesticide removal applications.

Preliminary trials conducted by the authors at a Thames Water site have demonstrated pesticide breakthrough for atrazine in 3.5 months, or 25,000 BV using GAC with an EBCT of 6 minutes. This indicates that GAC installed for taste and odour control has limited long term effect on pesticide levels, an observation subsequently confirmed from full-scale monitoring at Thames Water sites.

At present, the best data available to the authors suggest that GAC with an EBCT of 15 to 30 minutes will remove pesticide levels found in typical surface waters to below 0.1ug/l for a period of 6 to 24 months between regenerations. Long term trials are currently in progress at two Thames Water sites (one chemical coagulation and one slow sand filtration) to evaluate several of the available process options. Combinations of conventional treatment with GAC and ozone/GAC, using different types of GAC and EBCTs in the range 15-30 minutes, are being investigated.

At laboratory scale, Freundlich isotherm coefficients and results from accelerated column tests (ACTs) on micro GAC columns can be used to produce theoretical computer models of pesticide removal by longer contact time (10-30 minutes) GAC filter beds. Carbon suppliers, the UK WRc and US EPA have developed versions of these models, but little data have been published on model validation for multiple pesticides at typical concentrations in natural waters. Research is known to be developing in this area with the potential to save much site specific pilot testing if successful.[16]

Activated carbon for removal of THMs and precursors

Virgin activated carbon is widely reported[8,20] as capable of removing many of the organic carbon precursor compounds that can react with chlorine to produce THMs and other halogenated organic disinfection byproducts. For lowland river sources with a TOC concentration range of 2-6mg/l, GAC bed life is typically reported[20] as being between 10,000 and 15,000 BV for both prechlorinated and unchlorinated feed water. This equates to less than 3 months for an EBCT of 10 minutes.

Thames Water has installed GAC as a sand replacement in existing rapid gravity filters at the conventional chemical coagulation plants at Farmoor and Culham, to give an EBCT of 6 minutes for taste and odour control. Monitoring of THM levels confirmed the generally reported finding that short contact GAC alone is not an effective process for long term THM control or precursor removal.

Ozone with GAC : biological activated carbon

Replacement of pre-chlorination with pre-ozonation, or provision of an ozonation step prior to GAC is reported[8,18,19] to promote oxidation and biological removal of THM precursors, increasing GAC bed life significantly. The combined process has been referred to as biological activated carbon (BAC) and is common in France.

Thames Water has studied ozone/GAC and ozone/slow sand filtration processes for biological removal of THM precursors.[15,21] Results from a recent trial, comparing GAC and ozone/GAC are shown in Table 6.

Ozone/GAC appears to be a very effective process combination for controlling both pesticides and THM formation but some potential problem

Table 6   *Ozone / GAC and GAC for THM precursor removal*

OPERATING CONDITIONS			
	Raw Water TOC	(mg/l)	4 - 6
	Ozone Dose	(mg/l)	2 - 3
	Ozone Contact Time	(mins)	5
	GAC EBCT	(mins)	6
	THM Formation Conditions:		
	Chlorine contact time (t)	(mins)	30
	Chlorine residual at (t)	(mg/l)	0.5
RESULTS			
		Total THMs (ug/l)	Chloroform (ug/l)
Raw Water		30 - 50	15 - 30
GAC:	Weeks 1 - 5	2 - 10	1 - 5
	Weeks 10 - 30	15 - 25	10 - 15
Ozone/GAC:	Weeks 1 - 5	1 - 10	1 - 3
	Weeks 10 - 30	10 - 15	2 - 4

been identified.[19,22,23] Thames Water are undertaking pilot trials to investigate several areas:

* What are the optimal ozonation conditions and type of GAC?
* Is formation of brominated THMs enhanced by ozonation, and what are other ozonation by-products?
* Ozonation increases the biodegradable or assimilable organic carbon (AOC) in the water. Is this fully removed on biological activated carbon or could it pass into supply causing aftergrowth problems?
* Does the enhanced microbial activity in BAC result in shedding of excessive levels of heterotrophic bacteria?
* Can carbon fines laden with heterotrophic and pathogenic bacteria pass from the carbon bed into supply?

Removal of chlorinated solvents with GAC

GAC adsorption competes with air stripping as the only proven options for removing chlorinated solvents such as TCE and PCE from groundwaters. GAC does not have a high adsorption capacity for chlorinated solvents and requires frequent regeneration if TCE and PCE levels in source waters exceed 100ug/l.In some cases groundwaters contaminated with solvents also contain triazine herbicides, which cannot be removed by air stripping. GAC can be used for control of both types of contaminant.

Thames Water installed a 20 Mld GAC plant for solvent removal at a site in South East London during 1989, following sporadic appearances of TCE and PCE in the groundwater being pumped.

The literature[24] on adsorption of chlorinated solvents onto GAC provides isotherm data for single component systems, but little design data or full-scale operational experience. ACT data from microcolumns can be obtained from some carbon suppliers and WRc for use in computer modelling of removal efficiency and bed life.

Air stripping

Air stripping is not reported as a viable option for the removal of water soluble, non-volatile herbicides from water.

Stripping of preformed THMs which are volatile has been discussed in the literature[24] either by using conventional forced draught, countercurrent packed towers or by employing diffused air aeration. The authors are not aware of any full-scale plants operated by water utilities for this purpose, but believe that the food and drinks industry have utilised this process.[25]

The removal of volatile chlorinated organic solvents from groundwater has been the main application for air stripping processes,[24] with several plants currently operational in the UK. Typical designs utilise 6 to 10m high packed towers, operated with countercurrent water and air streams. Air to water ratios of 10 to 30:1 are reported.[24]

Main problem areas have been with calcium carbonate scale formation in hard water areas and a need for pH correction resulting from gas stripping of dissolved carbon dioxide. In some parts of the world, notably California, transfer of solvent pollution from the water phase to the atmosphere is not permissible. Gas phase GAC is then required in addition to the air stripping system, significantly adding to treatment costs.

Advanced oxidation processes

A number of studies in the US, Australia and France have indicated the potential for using 'Advanced Oxidation Processes' (AOPs) to completely destroy micropollutants such as pesticides and chlorinated solvents.[13,14,22,26] AOPs use combinations of oxidants, ultraviolet irradiation and catalysts to generate hydroxyl free radicals in water. These are highly reactive species capable of oxidising organics to carbon dioxide, water and mineral salts. Free radical scavengers, such as bicarbonate in hard water areas, can reduce the effectiveness of AOPs.

Key oxidant/UV combinations reported are:-
* Ozone + hydrogen peroxide (peroxone)
* Ozone + UV
* Hydrogen peroxide + UV
* UV + titanium dioxide catalyst

Peroxone and ozone/UV have received most attention, principally for destruction of TCE and PCE and for destruction of atrazine. Removals of 80 to 95 per cent have been reported for contact times of 2-10 minutes, ozone doses of 1-5 mg/l and hydrogen peroxide doses of 0.5-2mg/l.

Large-scale trials of ozone/hydrogen peroxide have been proposed in France and California for 1990 and 1991. Limited trials by the authors have been sufficiently encouraging for Thames Water to commence designs for a 5 Mld Peroxone trials plant, for completion in 1991.

The combination of UV and a semiconductor such as titanium dioxide is potentially attractive as it eliminates the need to generate ozone or purchase hydrogen peroxide. However, many process and scale-up problems remain to be solved for this option.

Membrane processes

Reverse osmosis (RO) has been shown to be effective at removing pesticides from water.[27] Very high operating pressures (10-70 atmospheres), long contact times and low throughputs per membrane do not provide encouraging economics for large-scale treatment. Nevertheless RO has been used for large-scale desalination and should not be disregarded for small supplies.

Lower pressure membranes (3-10 atmospheres) operating in the nanofiltration range offer better process economics for removal of THM precursors,[28] and several demonstration scale plants have been constructed worldwide. Thames Water is currently studying ultrafiltration, nanofiltration and RO, and monitoring TOC, THM precursor and pesticide removal performance.

Microbiological processes

There is a considerable body of literature on the microbiological degradation of pesticides, using either selected bacterial cultures or cell-free enzyme extracts.[29] However, this appears to relate to the use of microbiological methods for the detoxification of high concentration pesticide residues in soil, or in contaminated containers.

Little is published[6] on microbiological processes for removing pesticides and chlorinated solvents from natural waters, where these compounds would always be only a small fraction of the total carbon source available for microbial growth. Nevertheless, several research programmes are currently studying this area with conditioned or adapted bacteria at laboratory and pilot scale. The authors have found no evidence of atrazine removal across biological slow sand filters.

Novel absorbents

Synthetic macroreticular resins such as the Amberlite XAD series have been reported as suitable for removing pesticides from water at pilot scale. Chemical regeneration using an alcohol was reported, with overall process economics comparing favourably with GAC.[6] As yet the authors are not aware of any larger-scale trials of this process.

Conclusions

Many water utilities and national research organisations in the UK, Europe and North America are currently researching for solutions to the problems of pesticides, chlorinated solvents and chlorination by-products in drinking water.

Thames Water plc is supporting the national research programme being undertaken by WRc, and is also investing over £5 million in research, constructing a 5 Mld advanced water treatment demonstration plant for completion in 1991. It will include ozone, GAC and AOPs in conjunction with conventional treatment. This will be followed by a planned £200-£300 million capital investment in advanced water treatment processes by Thames Water in the 1990s.

The authors conclude that:
1. Pesticide and chlorinated solvent levels in source waters can exceed current UK standards.
2. Conventional water treatment processes do not effectively remove all pesticides and solvents.
3. Conventional chemical coagulation treatment and chlorination can form THM levels that exceed UK standards. Slow sand filtration and chloramination can meet current total THM standards. Chloroform and future THM standards are of concern.
4. Granular activated carbon, ozonation and air stripping offer advanced treatment options for the near future, though some process problems remain.
5. GAC can remove pesticides but not control THM formation. Ozonation is effective in reducing THM formation but will not remove all pesticides. Treatment process streams combining ozone and GAC may be the best solution.
6. Novel techniques involving advanced oxidation processes are close to full-scale introduction and may, if proven, offer economic and environmental advantages over GAC and air stripping. However, by-product studies are required.
7. Other novel processes using membranes, biological treatment and resin adsorbents are still at the bench and small pilot scale.

Acknowledgements

The authors wish to thank the Thames Water Engineering Director, W. Alexander, and Water and Environmental Science Director, P. McIntosh for permission to publish this paper. The authors also wish to acknowledge the contributions to this work made by Thames Water analytical scientists and researchers, and other colleagues. The views expressed are those of the authors.

References

1. British Medical Association. (1990) *Pesticides, Chemicals and Health*. BMA.
2. 'Statutory Instrument 1989 No. 1147, Water England and Wales'. *The Water Supply (Water Quality) Regulations 1989*. HMSO, 1989.
3. European Community. 'Council Directive relating to the Quality of Water intended for Human Consumption, 80/778/EEC'. *Official Journal*. 1980.
4. DoE Policy Letter WP/18/89.
5. Croll, B.T. (1988) 'Pesticides and Other Organic Chemicals'. IWEM Symposium on Catchment Quality Control, London.
6. Hart, J. and Jones, J.H. (1989) *Removal of Pesticides from Water: A Literature Survey*. WRc Report UM 1005.
7. Connorton, B. (1990) 'Groundwater Resources - A Thames Water Utilities Perspective', presented at the I.B.C. Conference: Water Availability Quality and Cost. London.
8. Walker, I., Shepherd, D.P., and Hyde, R.A. (1990) *Control of Chlorination Byproducts in Drinking Water from Lowland Supplies*. WRc Report UM 1118.
9. Gordon, G., Slootmaekers, B., Tachiyashiki, S. and Wood, D.W. (1990) 'Minimising Chlorite Ion and Chlorate Ion in Water Treated with Chlorine Dioxide'. *J.AWWA*.
10. Rachwal, A.J., Bauer, M.J. and West, J.T. (1988) 'Advanced Techiques for Upgrading Large Scale Slow Sand Filters', in *Slow Sand Filtration - Recent Developments in Water Treatment Technology*, Graham N.J.D. ed. Ellis Horwood.
11. Thames Water Internal Report.
12. Reynolds, G., Graham, N., Perry, R. and Rice, R.G. (1989)'Aqueous Ozonation of Pesticides: A Review'. *J. Ozone Science and Engineering*. Vol 11, 44.
13. Bouillot, P. and Poillard, H. (1990) 'Informative Note on Possibilities of Triazines Removal from Water'. CGE - Anjou Recherche internal paper.
14. Duguet, J.P., Bernazeau, F. and Mallevialle, J. (1990) 'Removal of Atrazine by Ozone and Ozone-Hydrogen Peroxide Combinations in Surface Water'. *J. Ozone Science and Engineering* Vol 12, 2.

15. Rachwal, A.J., Bauer, M.J. and Chipps, M.J. (1985) 'Ozone's Role in Biological Filtration Processes'. IOA Conference, Edmonton.
16. Hart, J., Sivil, D.C. and Carlile, P.R. (1990) *Pesticide Removal from Water - Bench Scale Studies*. WRc Report UM 1081.
17. Chemviron Carbon. Manufacturer's Literature for Filtrasorb 400 DS-A-02/88-12/p.2/Belgium.
18. Servais, P., Billen, G., Ventresque, C. and Bablon, G. (1989) 'Microbial Activity in Granular Activated Carbon Filters at the Choisy-Le-Roi Drinking Water Treatment Plant'. *J.AWWA*.
19. Prevost, M., Duchesne, D., Coallier, J., Desjardins, R. and Lafrance, P. (1989) 'Full Scale Evaluation of Biological Activated Carbon Filtration for the Treatment of Drinking Water'. AWWA WQ *Technical Conference Procedures*.
20. Lykins, B.W., Clark, R.M. and Adams, J.Q. (1988) 'Granular Activated Carbon for Controlling THMs'. *J.AWWA*.
21. Rachwal, A.J., Rodman, D.J. and West, J.T. (1984) 'Uprating and Upgrading of Slow Sand Filters by Pre-ozonation'. *Proceedings of Seminar on Ozone in Water Treatment Practice - The Future*. IWES WRc.
22. McGuire, M.J., Ferguson, D.W. and Aramith, J.T. (1989) 'Overview of Ozone Technology for Organics Control and Disinfection'. AWWA Water Quality Technology Conference - Seminar S.
23. McFetters, G.A., Camper, A.K., Davies, D.G., Broadway, S.L and LeChevallier, M.W. (1989) 'Microbiology of Granular Activated Carbon Used in the Treatment of Drinking Water, in *Biohazards of Drinking Water Treatment*. Ed. Larson R.A., Lewis Publishers.
24. Booker, N.A., Hart, J., Hyde, R.A. and Davies, G. (1988) *Removal of Volatile Organics from Groundwater*. WRc. Report 756;S.
25. Private communication from ETA - Air Stripping Plant Supplier. (1989).
26. Glase, W.H. and Kang, J.W. (1988) 'Advanced Oxidation processes for Treating Groundwater Contaminated with TCE and PCE: Laboratory Studies'. *J.AWWA*.
27. Baier, J.H., Lykins, B.W., Frank, C.A. and Kramer, S.J. (1987) 'Using Reverse Osmosis to Remove Agricultural Chemicals from Groundwater'. *J.AWWA* 79(8).
28. Watson, M.M., and Hornbur,g C.D. (1989) 'Low-Energy Membrane Nanofiltration for Removal of Colour, Organics and Hardness from Drinking Water Supplies'. *Desalination* 72 (1-2): pp 11-22.

29. Johnson, L.M. and Talbot, H.W. (1983) 'Detoxification of Pesticides by Microbial Enzymes'. *Experimentia* 39, Basle.